THE RIVER RUNS BLACK

THE RIVER RUNS BLACK

THE ENVIRONMENTAL CHALLENGE TO CHINA'S FUTURE

ELIZABETH C. ECONOMY

A COUNCIL ON FOREIGN RELATIONS BOOK

CORNELL UNIVERSITY PRESS
ITHACA & LONDON

The Council on Foreign Relations is dedicated to increasing America's understanding of the world and contributing ideas to U.S. foreign policy. The Council accomplishes this mainly by promoting constructive debates and discussions, clarifying world issues, and publishing *Foreign Affairs*, the leading journal on global issues. The Council is host to the widest possible range of views but an advocate of none, though its research fellows and Independent Task Forces do take policy positions. Please visit our website at www.cfr.org.

The Council takes no institutional position on policy issues and has no affiliation with the U.S. government. All statements of fact and expressions of opinion contained in all its publications are the sole responsibility of the author or authors.

Copyright © 2004 by Cornell University

All rights reserved. Except for brief quotations in a review, this book, or parts thereof, must not be reproduced in any form without permission in writing from the publisher. For information, address Cornell University Press, Sage House, 512 East State Street, Ithaca, New York 14850.

First published 2004 by Cornell University Press
First printing, Cornell paperbacks, 2005
Printed in the United States of America

Library of Congress Cataloging-in-Publication Data

Economy, Elizabeth C., 1962–
    The river runs black : the environmental challenge to China's future / Elizabeth C. Economy.
       p. cm.
Includes bibliographical references and index.
    ISBN 0-8014-4220-6 (cloth : alk. paper)
    ISBN 0-8014-8978-4 (pbk. : alk. paper)
    1. Environmental policy—China.   2. Economic development—Environmental aspects.   3. China—Economic conditions.   I. Title.
    HC430.E5E36 2004
    333.7'0951—dc22
                                        2003024994

Cornell University Press strives to use environmentally responsible suppliers and materials to the fullest extent possible in the publishing of its books. Such materials include vegetable-based, low-VOC inks and acid-free papers that are recycled, totally chlorine-free, or partly composed of nonwood fibers. For further information, visit our website at www.cornellpress.cornell.edu.

Cloth printing   10 9 8 7 6 5 4 3 2
Paperback printing   10 9 8 7 6 5 4 3 2 1

To my family

CONTENTS

Acknowledgments ix

CHAPTER ONE
The Death of the Huai River 1

CHAPTER TWO
A Legacy of Exploitation 27

CHAPTER THREE
The Economic Explosion and Its
Environmental Cost 59

CHAPTER FOUR
The Challenge of Greening China 91

CHAPTER FIVE
The New Politics of the Environment 129

CHAPTER SIX
The Devil at the Doorstep 177

CHAPTER SEVEN
Lessons from Abroad 221

CHAPTER EIGHT
Averting the Crisis 257

Notes 275
Index 327

ACKNOWLEDGMENTS

For more than two decades, I, along with the rest of the world, have watched as the Chinese people have transformed their country from a poverty-stricken nation into an economic powerhouse. Equally striking, however, has been the terrible price China's environment has paid for this impressive transformation. Today, the environment is beginning to exact its toll on the Chinese people, impinging on continued economic development, forcing large-scale migration, and inflicting significant harm on the public's health. My hope is that this book will provide support for the many in and outside China already engaged in developing new approaches to integrating economic development and environmental protection and will alert others to the possibility for change. It is a challenge that neither the Chinese people nor the rest of the world can afford to ignore.

In my study of China, I have received the unstinting support and wise counsel of many people. As a freshman at Swarthmore College, I received my introduction to China from a renowned scholar, Kenneth Lieberthal. Ken set a standard for rigorous analysis and scholarship that I have strived to meet ever since. Most important, while my initial academic and professional focus was on the Soviet Union, it was Ken who, on my return to school for graduate study at the University of Michigan, encouraged me to include China as a more central part of my work. Along with Ken, Harold Jacobson, Matthew Evangelista, William Zimmerman, and Thomas Princen all kept their office doors wide open, sharing generously their insights, time, and expertise in ways critical to the development of my scholarship.

For the past ten years, I have been a Fellow at the Council on Foreign Relations. During this time, I benefited immeasurably from the support and guidance of former Council president Leslie Gelb. Like all wise leaders, Les always challenged me to be better, without ever impinging on my freedom to set my own research agenda. My respect for his judgment, intellect, and wit is enormous, despite my occasional irreverence. Les has also created a dynamic community of Fellows at the Council, and my work has benefited in every way from my interaction with these talented scholars. David Victor and Adam Segal have been particularly generous in taking time from their own research to review portions of this manuscript. Two former directors of studies at the Council, Kenneth Keller and Lawrence Korb, have done much to make the Council's Studies department a collegial and supportive environment. And the current leadership of the Council, president Richard Haass, executive vice president Michael Peters, and director of studies James Lindsay, has supported me in all facets of my work on this book. The Council also provides the opportunity for its Fellows to vet their work before a group of distinguished outside experts. I was blessed to have Ambassador J. Stapleton Roy chair this group. He shared his vast experience and knowledge of China and welcomed participation from everyone, making the sessions lively and informative. Each of the group's members enriched my understanding of China and the environment and thus this book in important ways. This study group included Marcia Aronoff, Joanne Bauer, Frances Beinecke, Thomas Bernstein, Jan Berris, Carroll Bogert, Marcus Brauchli, Alan Brewster, Ralph Buultjens, John Chang, Jerome Cohen, Joel Cohen, Hart Fessenden, John Frankenstein, Jeffrey Glueck, Scott Greathead, Christopher Green, James Heimowitz, Richard Herold, John Holden, Francis James, Daniel Katz, Helena Kolenda, John Langlois, Herbert Levin, Kenneth Lieberthal, Zhi Lu, Jeanne Moore, Andrew Nathan, J. David Nelson, Sheridan Prasso, Theodore Roosevelt IV, Daniel Rosen, John Ryan, James Silkenat, Ann Brownell Sloane, Allan Song, Lawrence Sullivan, Stephen Swid, Jennifer Turner, David Victor, and Stephen Viederman. Irina Faskianos, vice president of the Council's National Program, also arranged two enormously helpful review sessions with Council members in San Francisco and Los Angeles that were chaired by William Reilly and Mathew Petersen, respectively.

At various points throughout my research on China and the environment, others too have provided invaluable assistance, intellectual guidance, and moral support. David Bachman, William Chandler, Karl Eikenberry, Eugene Matthews, Douglas Murray, Minxin Pei, Lester Ross, Vaclav Smil, Robert Socolow, and Karl Weber, in particular, stand out in this regard.

Many Chinese environmentalists, scientists, and officials have been extraordinarily patient in helping me navigate the world of China's environmental politics and have shared their expertise with me repeatedly for more than ten years. Many have become close friends, and, while they remain unnamed, I owe them a special debt of gratitude.

Support for the book has come most significantly from the Smith Richardson Foundation, where Marin Strmecki, Samantha Ravich, and her successor Allan Song were early supporters of my interest in exploring the broader political and economic implications of China's environmental practices. T. C. Hsu and Florence Davis through their leadership of the Starr Foundation also contributed travel and research funds. The innovative international peace and security studies fellowship program under the auspices of the John D. and Catherine T. MacArthur Foundation and the Social Science Research Council provided me with my first opportunity to learn about environmental issues as a dissertation fellow at Princeton University, and later supported my work during my first years at the Council on Foreign Relations. I have learned that economic support must begin with a shared appreciation of an idea, and I am fortunate that all of the above have such an interest in helping others transform ideas into books.

Throughout my tenure at the Council, I have been blessed by an outstanding group of research associates, including Nancy Yao and Vanessa Guest. I am delighted to watch as each of these talented individuals achieves success in her career. I also had the advantage of superb research assistance from Eric Aldrich, who during the three key years of research and writing for this book made it his own mission. He spent countless hours on research and expertly managed the study group process, all with unflagging enthusiasm. He is already making his mark as an expert on acquired immunodeficiency syndrome (AIDS) and other communicable diseases in China. Laura Geller picked up where Eric left off and has made an invaluable contribution to the book in its final stages.

I thank Roger Haydon at Cornell University Press for his enthusiasm for my manuscript and willingness to move quickly to bring it to fruition. Karen Hwa and Cathi Reinfelder have also exercised extraordinary patience in overseeing the editorial process. To the two anonymous reviewers who read the entire manuscript with such care, I offer my gratitude for pointing out all the ways in which I could make the book more useful to a wider audience.

To have so many to thank for their support in writing this book is my good fortune. To have the privilege of knowing someone who contributed as much to my learning of and passion for the study of China as Michel Oksenberg is my miracle. Mike's intuitive understanding of China made him a preeminent China scholar. For me, that intuition, zest for life, generosity of spirit, and unbounded optimism made him invaluable as an adviser and friend. Tragically Mike passed away in February of 2000, and it is my great sorrow that he is not here for me to thank. Throughout the writing of this book, his memory has sustained me, and wherever my future work takes me, my fifteen years as Mike's student, collaborator, and friend will continue to inspire me.

No amount of counsel and guidance during my university years and professional career could substitute for the love and support of my family. My parents, James and Anastasia, encouraged me to be curious, questioning, and independent, unwittingly preparing me well for the years of research to come. Through their own example, they also instilled in me and my siblings, Peter, Katherine, and Melissa, a desire to think about the world beyond and to find a way to contribute to society. My research on China and this book in particular represent one small effort to fulfill such aspirations. In my family, I have been doubly fortunate to have been married for the past ten years to David Wah. His humor, love, and editing skills have been essential to my ability to produce this book as well as three energetic and engaging children, Alexander, Nicholas, and Eleni. Without their enthusiasm for my work and patience for my frequent travels, this book could never have been completed. Above all, this book is my thanks to my family.

Finally, let me note that all responsibility for any errors in this book is mine alone and apologize to anyone whom I have unknowingly excluded from these acknowledgments.

ELIZABETH C. ECONOMY

New York City

THE RIVER RUNS BLACK

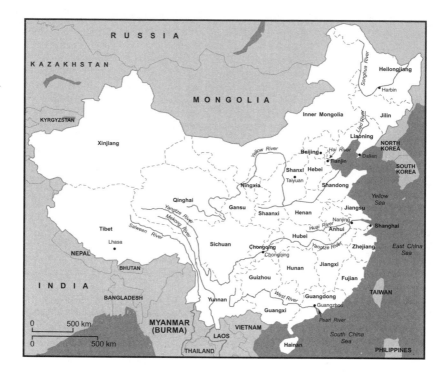

THE DEATH OF THE HUAI RIVER

In late July 2001, the fertile Huai River Valley—China's breadbasket—was the site of an environmental disaster. Heavy rains flooded the river's tributaries, flushing more than 38 billion gallons of highly polluted water into the Huai.[1] Downstream, in Anhui Province, the river water was thick with garbage, yellow foam, and dead fish.[2] Although the authorities quickly proclaimed the situation under control, the incident represented a stunning failure for China's leadership. Only seven months earlier, the government had proclaimed its success in cleaning up the Huai. A six-year campaign to rid the region of polluting factories that dumped their wastewater into the river had ostensibly raised the quality of the water in the river and its more than one hundred tributaries to the point that people could once again fish, irrigate their crops, and even drink from the river.

The story of the Huai River over the past five decades epitomizes the saga of environmental change in China. It's a paradoxical tale, one that holds out the promise of significant change in the future, while exposing the failures of China's current environmental practices, many of which are rooted in centuries-old traditions.

The Huai River Valley, including Anhui, Jiangsu, Shandong, and

Henan Provinces, is a fertile region in eastern China. It is roughly the size of England with a population of over 150 million people, all of whom depend on the Huai for their water supply. The river originates in Henan's Tongbai Mountain and flows east for over six hundred miles through Henan, Anhui, and Jiangsu Provinces before flowing into the Yangtze River.

The Huai River Valley is a relatively prosperous region, with average per capita incomes in 2002 ranging from just under $800 in Anhui to almost $1,760 in Jiangsu.[3] Long known for its rich supply of grain, cotton, oil, and fish, the river basin has over the past two decades become home to tens of thousands of small factories. Paper and pulp mills, chemical factories, and dyeing and tanning plants, employing anywhere from ten to several thousand people, have sprouted all along the banks of the river and its tributaries, driving much of the economic dynamism of the region and the nation. They have also freely dumped their waste into the river, making the Huai China's third most polluted river system.[4]

The Huai River boasts a dramatic and tumultuous history. In 1950, disastrous flooding prompted Mao Zedong to create the Huai River Basin Commission. As part of Mao's campaign to control rivers, the Commission commandeered tens of millions of Chinese to construct no fewer than 195 dams along the river.[5] In August 1975, two of the largest dams, Shimantan and Banqiao, collapsed, killing an estimated 230,000 people.[6]

The dams have also contributed to the numerous pollution disasters that have plagued the Huai River for more than two decades. Local officials upstream have repeatedly opened the sluice gates of the dams, releasing polluted water that has poisoned crops and fish downstream thereby ruining local farms and fisheries. The problem is compounded by the roughly four thousand reservoirs constructed along the river, which limit the river's capacity to dilute the pollutants.[7] In many stretches of the river, the water is unfit for drinking.

Despite relatively high average annual rainfall of thirty-four inches, many parts of the river basin are also prone to drought,[8] worsening the concentration of pollutants. Along some parts of the river, people have long recorded higher than normal rates of cancer and birth defects. According to one estimate, the death rate along one stretch of the main river is one third above the provincial average, and the cancer rates are twice the provincial average.[9] A report on the region in the late 1990s also notes that "for years no boy from [certain villages in] the Huai River area has been healthy enough to pass the physical examination required to enter the army."[10]

The Chinese leadership has not been unaware of the river's growing pollution. After the first pollution disaster in 1974, the Chinese leadership in Beijing established the Huai River Valley Bureau of Water Resources Protection and the Huai River Conservancy Commission of the Ministry of Water Resources. However, these offices had no funding and no real authority, and as the region developed economically, the environment deteriorated rapidly. As one official from China's central environmental agency, the National Environmental Protection Agency (NEPA) described the situation, "Economic development had just occurred blindly."[11] In 1988, Beijing established a central government Leading Group on Water Resources Protection for the Huai River Basin, raising the profile of the problem at least bureaucratically. But by 1990, to cut costs and realize greater profits, fewer than half of the factories in the valley were operating their waste disposal systems, and only 25 percent of treated wastewater met state standards. Sensing impending disaster, the Bureau of Water Resources Protection pressed local officials to close down or retrofit some of the most egregious polluters. Some were closed, but others quickly opened in their stead.[12]

Moreover, the four provinces dependent on the Huai were incapable of coordinating a policy to address the problem. By 1993, the director of the Bureau of Water Resources Protection was complaining to no avail about the rising number of interprovincial disputes over the Huai and the lack of any authority capable of resolving them. In May 1994, Beijing responded—at last—to the warnings of the local environmental officials and, perhaps, to the growing social unrest in the region.[13] The country's top environmental oversight body, the State Environmental Protection Commission (SEPC), under the auspices of the State Council, convened a meeting in Anhui Province to discuss the problem of provincial cooperation on environmental issues. But the meeting produced no tangible change in policy or practice.

Just two months later, a number of factories along the Huai emptied their waste tanks directly into the river, producing a toxic mix of ammonia and nitrogen compounds, potassium permanganate, and phenols, contaminating the middle and lower reaches of the river. The water turned black, and factories were forced to close down. Fisheries were destroyed, almost 26 million pounds of fish were killed, and thousands of people were treated for dysentery, diarrhea, and vomiting.[14] In the immediate aftermath, local authorities in one town (Bengbu, Anhui) released polluted water that had been pent up by their local dam. During the following two weeks, more than 52 billion gallons of polluted river water were released. As the media flocked to the scene, local officials attempted to hide the extent of the disaster; in response, villagers pelted the officials with eggs.[15]

The central leadership immediately dispatched an investigation team, headed by Song Jian, the highly respected chairman of the SEPC. During the team's visit, a peasant offered Song a glass of the river water to drink. After sipping from the cup, Song invited the provincial and local officials to finish the glass, informing them simultaneously that if they did not clean up the river, they would be fired.[16] Some action followed. Reportedly nearly one thousand factories were closed down or relocated; others received a three-year grace period until 1997 to improve their environmental practices. Premier Li Peng announced a two-stage campaign: first to halt all industrial waste pollution by 1997, and second to have the river run clear by 2000.

To accomplish this goal, the State Council established an interministerial subcommission including the Ministry of Water Resources, the State Planning Commission, and the Ministry of Finance, among others.[17] Director of NEPA Xie Zhenhua announced specific steps to improve the situation; nineteen industrial firms in four provinces were given specific targets to meet. In 1996, hundreds of polluting factories along the Huai were closed.

Yet the imperative of economic development continued to overwhelm environmental concerns. As Cang Yuxiang, a member of NEPA's Pollution Control Division, reported in 1997:

> They [local industries and regions] pay attention only to short-term interests to the detriment of the long-term, or their own area

to the detriment of the entire river basin and, disregarding the harm done to others, allow large amounts of waste water to flow to other areas. . . . Towns and villages continue to blindly build small paper mills, dye works, tanneries, and chemical plants with crude equipment, despite the government already having temporarily closed down some such 5000 factories during the cleanup process. . . . Most of the water in the river system is rated at levels 4 and 5 (on a scale ranging from 1 to 5, the higher numbers indicating greater pollution) with some of the tributaries even failing to meet the number five standards and in the dry season basically becoming wastewater sewers.[18]

There were additional reports that the equipment from the shuttered enterprises was being sold to other factories,[19] and the U.S. Embassy reported that within two years, 40 percent of the closed factories had reopened.

Still, the government pressed on with the second stage of its campaign. On January 1, 1998, the Chinese leadership launched the three-year Zero Hour Operation (*Lingdian Xingdong*) to clean up the Huai River. In short order, local officials closed down 35 factories and halted production at an additional 198 plants.[20] Yet researchers at the Chinese Academy of Social Sciences Rural Development Institute voiced their doubts concerning the future efficacy of the campaign:

It is very unlikely that very serious regional pollution can be solved through the Zero Hour Operation. . . . Before the Zero Hour Operation arrives, many polluters will suspend discharging waste until the NEPA finishes its examinations. . . . The situation is unlikely to be reversed in the near future. It is difficult for the state to collect pollution discharge fees from small and scattered factories. They don't produce any proper accounts and no one can be sure about their finances.[21]

In discussing the problem, one of China's leading environmentalists at the time reported on the tactics used by factories along the

Huai for avoiding closure. He commented that some of the small paper and pulp mills banded together to form large plants in order to evade government regulations on the size of the factories; other factories closed down during the day but operated at night. Moreover, local governments, fearful of economic loss, exerted enormous pressure on environmental protection officials not to pursue the campaign aggressively. The Chinese investigative television program *Newsprobe* paid two visits to the region in 2000, first to report on the efforts to clean up the river and then to assess their success. Not surprisingly, during the second visit, *Newsprobe* discovered that many factories that should have been closed were still operating.[22]

Reports of problems multiplied. In January 1998, just after the initial Zero Hour push, the water in Xiuzhou, an industrial city downstream on one of the Huai's tributaries, turned black, the fish died, and the residents were without running water for two weeks. (A few days earlier, the government had announced that the Huai was well on its way to "environmental resurrection.")[23] One resident complained to a foreign journalist, "The water here is always polluted. It's really serious. We can't even wash our clothes in this water. And when we tried to give it to the pigs, the pigs refused to eat for a couple of days."[24]

A second especially egregious case of continued pollution of the Huai involved Fuyang in Anhui Province, whose wastewater flows into the Huai. In 1999, Fuyang was designated one of China's ten "Clean Industry Cities." Yet in May 2000, ten residents of Fuyang collapsed along the city's main sewage outlet known as the "Seven-Li Trench." Six of the ten died from exposure to the contaminated water. While the Fuyang government had already spent almost $600,000 on a system of sluice gates designed to control the flow of wastewater, they refused to spend the $30 million needed to clean up the wastewater that flowed through the gates into the Huai, arguing that attacking the problem aggressively would require the closure of much of the city's industrial base.[25] The residents of Fuyang were well aware of the dangers of the polluted water; when the sluice gate emptying waste from the city's main industrial area was open, the water turned black and a foul smell and haze emanated from it. Although the local villagers had complained publicly about the Seven-Li Trench, their complaints probably died with unsympathetic local officials. A television news team that reported on Fuyang in 1999, for example, edited a farmer's criticism to sound like praise. It emerged that Fuyang's designation as a "Clean Industry City" was the result of a deception common to many cities: Local factory officials were always warned before environmental inspectors arrived, and they ensured that their factories passed the inspections by shutting off waste outlets and flushing systems with clean running water.[26]

Fuyang now faces severe water shortages. The local groundwater is so polluted that the city has been forced to dig wells deeper and deeper, causing severe subsidence. According to one report, "[U]nderground water [is now] completely drained."[27]

In 1999 and 2000, the Huai ran dry for the first time in twenty years.[28] Shipping came to a halt as boats were left "high and dry in the mud."[29] In Hongze Lake downstream, residents "witnessed the ghostly walls of Sizhou, a city submerged by floods 300 years ago, emerging into the light."[30] The local economy was hit hard with crops ruined and thousands of tons of fish dead.

Despite such reports, in January 2001, Xie Zhenhua, the head of China's State Environmental Protection Administration (until March 1998, the National Environmental Protection Agency), stepped forward to claim that the water quality on the Huai River had reached the national grade three standard—suitable for drinking and fishing—and that 70 percent of the major tributaries had reached grade four—suitable for industry and agriculture.[31] The *People's Daily*, China's foremost government newspaper, claimed that the kind of cleanup that had taken other countries twenty years to accomplish had been achieved by the Chinese in just a few years.

In the past, such assertions might have passed unchallenged. No more. On January 18, the *Worker's Daily* printed a front-page article asserting that the government's campaign had failed. Su Kiasheng, professor at Huainan Industrial College and vice-chairman of Anhui's People's Political Consultative Conference, a reform-oriented advisory body within the provincial government, claimed that the water was still seriously polluted and far from achieving the State Council's standards. Su also asserted that the discharge of pollutants in 2000 was actually double the government's targets and that the water quality of the Huai in Anhui was grade five—unsuitable even for irrigation: "To meet the State Council's targets, some provinces gave false figures and made false reports to the central government. As a matter of fact, the pollution has not changed much, although some of the smaller factories have been ordered to halt production." Professor Su further pointed out that according to the government's plan there were to be fifty-two water-treatment plants along the river; instead there were only six, with a few more under construction.[32]

As surprising as Su's willingness to challenge the official line was the willingness of the *Worker's Daily*, as well as the Shanghai bureau of the *People's Daily*, to publish articles in support of Su's story. Moreover, a letter to the editor pointing out the discrepancy between the *People's Daily* and the *Worker's Daily* was published in a third newspaper, *China Youth Daily*.[33]

Su's analysis was soon borne out. In summer 2001, torrential rains caused yet another pollution crisis on the Huai, pushing 144 million cubic meters (38 billion gallons) of polluted water, which had been stored behind water locks in the upper reaches of the river, downstream, killing fish and plant life and threatening drinking water. The central government reacted quickly, ordering one hundred polluting firms to slow down or stop production and sending an inspection team to the region.[34] In 2002, the World Bank also signed on to assist in the development of several municipal and enterprise wastewater treatment plants in Anhui and Shandong Provinces to reduce water pollution in the Huai River Basin.[35]

Still, improvement in the Huai's water quality remains a distant hope. Water quality reports by China's National Environmental Monitoring Center in 2003 reported that at the three monitoring sites along the mainstream of the Huai River, water quality was either grade four or five—unsuitable for drinking, fishing, or, in the worst case, agriculture and industry.[36] Moreover, the costs of pollution to the Huai region will only increase. Cleanup costs are now estimated at more than $100 billion, and the number is rising.[37] The region is bracing for a huge, expensive river diversion project to bring water from the Yangtze to the Huai, a project that officials hope will be largely complete by 2013. As for the economic toll of illness and death, it is incalculable.

Beyond the Huai

Sadly, the saga of the Huai River Valley is typical of today's China. Throughout the country, centuries of rampant, sometimes willful, destruction of the environment have produced environmental disasters.

- *Flooding.* In 1998, the Yangtze River flooded, killing more than three thousand people, destroying five million homes, and inundating fifty-two million acres of land. The economic losses were estimated at more than $20 billion. The culprit: two decades of rampant deforestation and destruction of wetlands.

- *Desertification.* Desert, already covering one-quarter of the country, continues to spread, with the pace of desertification having doubled since the 1970s. Efforts to stem the advance of the desert through afforestation and development of grassland have had limited success. Since 1998, northern China, including the capital of Beijing, has been shrouded in dust for days at a time. Year by year, this dust has traveled increasingly far afield, darkening the skies of Japan and Korea and even a wide swath of the United States.

- *Water scarcity.* Water scarcity born of changing ecosystems, skyrocketing demand, increasing pollution, and few conservation efforts has limited access to water for more than 60 million people, and almost ten times that number drink contaminated water daily. As a result, factories are closing, tens of millions of people are leaving their land, public health is endangered, and the government faces a future of expensive river diversion and pollution remediation projects, even in areas once rich in water resources.

- *Dwindling forest resources.* China's forest resources rank among the lowest in the world. Demand for furniture, chopsticks, and paper has driven an increasingly profitable, environmentally devastating illegal logging trade. By the mid-1990s, half of China's forest bureaus reported that trees were being felled at an unsustainable rate, and 20 percent had already exhausted their reserves. China's Sichuan Province—home to the famed pandas—now possesses less than one-tenth of its original forests. Loss of biodiversity, climatic change, desertification, and soil erosion are all on the rise as a result.

- *Population growth.* Underpinning much of China's resource degradation has been the continued pressure of a burgeoning population. China's top leader until March 2003, Jiang Zemin, called the country's population size his biggest problem, particularly in the countryside, where many routinely defy the coercive one-child policy. Even if the 2000 official census is accurate, China's 1.295 billion people exceed by almost 300 million the goal set by China's leaders a decade earlier.

At first glance, China's story appears to be a classic tale of economic development run amok: As China has moved to a market economy, freeing its economic actors to make money and exploit the country's natural resources without penalty, it now confronts an environmental crisis. Such an explanation fits squarely within the logic of thinkers such as Karl Polanyi, who predicted that the unfettered market would yield only negative consequences for the environment.[38] Others have argued convincingly, too, that China's large population and rapidly increasing integration into the international economy help explain why China now confronts such a substantial array of environmental challenges.[39] Yet, equally potent arguments are offered by others to suggest that economic development, population, and trade benefit a country's environment. Both sides of these long-standing debates over the relationship between economic development, population, and trade, on the one hand, and the environment, on the other, provide important clues for understanding China's environmental trajectory.

The Broader Debates

Whether economic development provides opportunities or challenges for environmental protection provokes sharp debate. Some

believe that development is necessarily harmful to the environment, demanding the extraction and consumption of natural resources, such as timber, minerals, oil, and water. Studies of developing countries in Southeast Asia, including Malaysia, Indonesia, and
the Philippines, offer some striking examples of how such exploitation of natural resources has contributed to devastating levels of deforestation, soil erosion, and desertification.[40] Proponents of this
line of thought also point out that levels of pollution rise as economic development proceeds. Economist Asayehgn Desta, for example, notes that in China, the development of rural areas has contributed to alarming levels of municipal waste and pollution from
increased pesticide and fertilizer use.[41] Underlying the problem, according to many theorists within this school of thought (such as
Polanyi), is that the market does not properly account for either the
use or the abuse of environmental and natural resources. So, as the
market and economy continue to develop, this accounting failure
becomes magnified and environmental problems increase.[42]

Others, however, counter that economic development brings important transformations within society that enhance environmental
protection, through value change, technological progress, and improved state capacity. As political scientist Ronald Inglehart has discovered through decades of survey research, as incomes grow and
poverty is reduced, educational standards rise and values such as environmental protection become more prevalent within society such
that demands for a better environment grow.[43] In addition, economic development contributes to broader changes within the economy that can have a positive impact on the environment, such as
growing access to technology that is environmentally friendly. As
noted international economist Jagdish Bhagwati has argued, economic growth provides governments with the ability to tax and raise
resources for a variety of social welfare goods, including environmental protection.[44]

This is not an either–or debate. Geographer Vaclav Smil suggests a
middle ground for China: "There are no solutions within China's
economic, technical, and manpower reach that could halt and reverse these degradative trends—not only during the 1990s but also
during the first decade of the new century." At the same time, he

notes that technical innovation, among other policy strategies, can reduce the future rates of China's environmental impacts from economic development.[45]

A second, related debate focuses on whether trade and foreign investment exert a negative or positive impact on a country's environmental situation. One commonly heard line of thought is that the comparative advantage in global trade for developing countries lies in the exploitation of their resources; therefore the impact on the environment is likely to be negative. In addition, trade, like the domestic market, does not take into account environmental costs and therefore further impedes a country's ability to protect the environment.[46] Moreover, competition for foreign investment may lead countries to specialize in pollution-intensive industries,[47] providing opportunities for other countries to offload their most polluting industries. As Yok-shiu Lee and Alvin So argue in their study of environmental movements in Asia, many multinationals transfer "substandard industrial plants and hazardous production processes" to Southeast Asia in order to avoid the health and pollution standards of their home countries. They note: "To a great extent, most of these trade- and investment-induced environmental degradation and pollution problems in Southeast Asia can be traced to multinational corporations that are either exploiting raw materials in or have shifted their industrial production operations to the region."[48] Finally, while some analysts acknowledge that trade-created wealth allows for social conditions to improve, they further point out that national governments will try to protect national industries. This in turn will contribute to inefficiencies in domestic production with negative environmental impacts.[49]

Yet, others have argued equally vociferously that economic development and integration into the international economy positively affect environmental protection. As Jose Furtado and Tamara Belt suggest, trade stimulates economic growth, optimizes efficiency of resource use, increases the standard of living and thus the demand for environmental services, and results in more funding of institutions for environmental protection.[50] Similarly, as John Audley argues in his study *Green Politics and Global Trade*, free trade simultaneously contributes to domestic economic growth, which will alleviate poverty and allow governments to spend more protecting the environment, and provides greater access to environmental technologies and less polluting industries.[51]

The third broad debate—perhaps the most contentious one—concerns how population affects the environment. It is encapsulated well by John Carey in his article, "Will Saving People Save Our Planet?":

> In particular, the question of just how much population control is necessary for sustainable development provokes a very nasty debate. On [one] side are the Malthusian pessimists like Stanford's Paul Ehrlich, who see teeming people as a malignant cancer that will grow to kill its planetary host. Unless the population bomb is defused, they say, no amount of development can save our priceless natural heritage. Nonsense, scoff the technological Panglossians, such as Ester Boserup, a Danish specialist in international development. They believe the additional people simply drive the engine of progress faster, raising standards of living for everyone. What keeps the disputes boiling is that plenty of evidence supports both positions.[52]

As Desta elaborates on the Malthusian perspective, "One consequence of population growth is the need for more land both for cultivation and for residential construction, leading to an increase in environmental deterioration. Similarly, population growth can lead to the decrease of per capita agricultural land, soil productivity, and a decline in nonrenewable resources such as coal, oil, and metal ores."[53] Many Chinese analysts, too, are among the strongest supporters of the necessity of controlling population to protect the environment:[54] "The Chinese population has long become [sic] a pressure on the resources in China. The overload, the excessive exploitation of the resources and the borrowing of the portion of resources due to the future generation have constituted a constant threat to the material base for the survival and development of the Chinese nation."[55]

Others, however, compellingly argue the benefits of population growth for the environment; namely, that population growth can drive technological progress either because of resource scarcity[56] or because population growth means more people to "drive the engine of progress faster"[57] and increases the chance that scientists will make discoveries that will improve the human condition in the long run.[58]

As Carey's comments indicate, these debates are kept alive because evidence exists to support both sides of each debate. As such, they do not provide a road map for understanding China's environmental trajectory. Instead, they offer some guidance in understanding both the environmental challenges and opportunities that China confronts as it develops economically, integrates into the international economy, and negotiates a burgeoning population.

Important, too, is what these debates do *not* address. In their purest form, they resist a discussion of the role of political institutions and politics in shaping a country's environmental and developmental pathway. But as many others have noted, excluding the political variable in these debates misstates the problem.[59] As China's story unfolds, politics—in a variety of forms—emerge as critical:

- Who are the key actors and what is their relative power?
- How are resources allocated to environmental protection?
- How is environmental policy formulated and implemented?
- What incentives do or do not exist for government, business, and society in China to advance goals of environmental protection?

Answering these questions helps explain why and under what circumstances factors such as economic development, population growth, and foreign trade and investment exert a positive or negative impact on China's environment. They are the key to understanding how China has arrived at its current environmental challenge and what this challenge portends for the country's future.

The Contemporary Political and Economic Context

As we will see, China's current environmental situation is the result not only of policy choices made today but also of attitudes, approaches, and institutions that have evolved over centuries. Yet, today, China's environmental practices are overwhelmingly shaped by the dramatic process of economic and political reform that has been transforming the country since the late 1970s.

The China inherited by Deng Xiaoping and his supporters twenty-five years ago reflected decades, if not centuries, of political and economic upheaval. Over two thousand years, dynasties rose and fell, typically amid great turmoil. China's first experiment with a republican form of government, beginning in 1912, soon fell victim to both internal and external challenges to its legitimacy. More than three decades of turmoil, civil war, and Japanese military aggression ensued, and General Chiang Kai-shek's efforts to unify China under his own brutal and corrupt leadership during this time ultimately collapsed at the hands of Mao Zedong and China's communists in 1949.

Mao moved quickly to establish an extensive, functioning bureaucracy and centrally planned economy modeled generally on that of the Soviet Union. Yet his pursuit of "continuous revolution" wreaked havoc on China's political and economic system. He instigated mass campaigns, such as the anti-rightist campaign and the Cultural Revolution, to route out those perceived to be not sufficiently revolutionary in their thinking; his quest for a communist utopia led to the Great Leap Forward, with its wildly unrealistic targets for steel and grain output. Together, these campaigns isolated China from the rest of the world and left tens of millions of Chinese dead or imprisoned and the Chinese economy in shambles.

By the time of Mao's death in 1976, the Chinese leadership had just begun the process of recovery. In 1971, the leadership took the first steps toward rejoining the international community by claiming the China seat in the United Nations (UN) from Taiwan and reestablishing relations with the United States, among other measures. In 1975, Premier Zhou Enlai enunciated the Four Modernizations (agriculture, industry, science and technology, and military) to revitalize China's economy and society.

With attainment of the Four Modernizations as their overarching objective, Deng Xiaoping and his supporters initiated a wholesale reform of the country's economic and political system. In the early

1980s, they began to relax the tight state control that, in one way or another, had defined China's economic and political situation since 1950.[60]

In the economic realm, this signaled the beginning of a transition from a state-directed, command economy to a more market-based economy. Beijing devolved significant economic authority to provincial and local officials, removing political constraints on their economic activities and diminishing Beijing's own ability to influence the development and outcome of these activities. China also began to invite participation from the international community in China's economic development through foreign direct investment and trade. By the mid-to late 1990s, the state had begun in earnest to dismantle the system of state-owned enterprises that had been the foundation of the urban economy, to encourage the expansion of private and co-operative ventures, and to energize the rural economy through the development of smaller-scale township and village enterprises.

In the political sphere, a similar transformation of institutions and authority took place, marked by four distinct processes of reform: (1) the highly personalized system of governance was transformed into a more institutionalized system with a codified system of laws; (2) significant political authority was devolved from center to local officials; (3) China embraced technological assistance, policy advice, and financial support from the international community; and, perhaps most dramatic, (4) as the government retreated from the market, it also retreated from its traditional role as social welfare provider, encouraging private, nonstate actors to fill the gap in areas such as education, medical care, and environmental protection.

China's reforms have significantly diminished the role of the state and encouraged newfound reliance on the market and the private sector to meet the economic and social welfare needs of society. They have also introduced new actors, institutions, and ideas into the political system. At the same time, as discussed below, the reforms have reinforced some of China's traditional policy attitudes and methods, sometimes in surprising ways. China's current environmental challenges, as well as its approach to environmental protection, reflect this rich, sometimes complicated mix of deeply rooted links to the past and dynamic new pressures and opportunities of today.

A Legacy of Devastation

The roots of China's current environmental crisis run deep. Through the centuries, the relentless drive of China's leaders to amass power, consolidate territory, develop the economy, and support a burgeoning population led to the plundering of forests and mineral resources, poorly conceived river diversion and water management projects, and intensive farming that degraded the land. As chapter 2 describes, this exploitation of natural resources, in turn, contributed significantly to the wars, famines, and natural disasters that plagued China through the centuries. The result was a continuous cycle of economic development, environmental degradation, social dislocation, turmoil, and often violent political change.

Furthermore, China lacked any compelling ethos of conservation. Rather, attitudes, institutions, and policies evolved from traditional folk understandings and philosophical thought, such as Confucianism, which most often promoted man's need to use nature for his own benefit.

As a result, little effort was made to develop institutions to protect the environment. China's leaders from the emperors to Mao Zedong relied on a highly personal system of moral suasion with few environmental regulations and no codified environmental laws. During the Qing dynasty, great effort was expended to develop a system of detailed laws, but it was not sustained after the dynasty fell. China's system of environmental protection was also highly decentralized, relying on individual regional and local officials to safeguard the environment within their jurisdiction.

China's leadership typically approached the environment through frequent campaigns and mass mobilization efforts for large-scale infrastructure projects, such as dams or river diversions that wreaked havoc with local ecosystems and were undertaken with little con-

sideration for the actual environmental and scientific factors necessary to achieve success. The campaign to clean up the Huai River, in
which the central government set unrealistic targets for pollution
control and failed to follow through with appropriate incentives or
disincentives for local government officials and businesses to
change their behavior, is emblematic of how deeply rooted the campaign mentality remains in China's political culture (as discussed in
chapters 4 through 6).

History therefore offered little environmental wisdom to China's
post-Mao reformers. As economic reform took hold, China experienced dramatic average annual growth rates exceeding 8 percent,
the elevation of hundreds of millions of Chinese out of poverty,
and, eventually, the transformation of China into a global economic powerhouse. No effort, however, was made to account for
the costs reform levied on the environment, and natural resources
remained priced far below their replacement costs in order to encourage continued rapid economic development. China's leaders
were well aware that they were trading environmental health for
economic growth. The maxim "First development, then environment," was a common refrain throughout the 1980s and much of
the 1990s.

Chapter 3 illustrates the enormous toll the reform period has
taken on the environment. Rates of air and water pollution have
skyrocketed. By 2002, China had become home to six of the ten
most polluted cities in the world. Acid rain now affects about one-
third of China's territory, including approximately one-third of its
farmland. More than 75 percent of the water in rivers flowing
through China's urban areas is unsuitable for drinking or fishing.
Desert now covers 25 percent of China's territory, and deforestation
and grassland degradation continue largely unabated.

The leadership has also begun to witness the broader social and
economic costs of its environmental failure. Over the next decade or
two, Beijing anticipates the migration of 30–40 million Chinese—
some voluntary, many forced—as a result of depleted or degraded resources. If not managed properly, the combination of migrant laborers and unemployed workers could trigger serious conflict in urban
areas, as has already happened in some cities. Managing this process

of migration and urbanization successfully will challenge the Party's ingenuity and organization, not to mention its financial resources.

The economic costs of environmental degradation and pollution are also dramatic. Already they are the equivalent of 8–12 percent of China's annual gross domestic product (GDP).

In addition, pollution-related illnesses are soaring. There have been serious outbreaks of waterborne disease, as well as long-term health problems in riverside communities reflected in rising rates of spontaneous abortion, birth defects, and premature death. Air pollution alone, primarily from coal burning, is responsible for over 300,000 premature deaths per year.[61]

Perhaps most threatening to the Chinese leaders, protests over polluted water, damaged crops, air pollution, and forced resettlement contribute to the increasingly pervasive social unrest already confronting them. Top leaders have commented on the danger that environmental protest poses to the authority of the Communist Party and the stability of the state, noting that it is one of the four most important sources of social unrest in the country.

At the same time, China's leaders are grappling with the legacy of the past as population and national security concerns continue to shape environmental practices. While an aggressive family planning effort has cut the birthrate in half, there are wide regional discrepancies, with important economic and environmental implications. And, even as national security concerns over issues such as civil war have diminished, others have emerged. In the mid-1990s, declining grain yields and growing reliance on international grain markets prompted concerns over China's food security, resulting in yet another massive land reclamation and grain-growing campaign to ensure grain self-sufficiency. Today, the Chinese leadership has turned its attention to China's west, where it confronts a combination of discontent among minority populations and lack of economic development. In response, China's leaders have initiated a massive economic development campaign to raise the standard of living and to secure China's borders from separatist movements—at great risk to the environment.

Yet, China's reforms have also changed the political landscape for environmental protection in more positive ways. As chapter 4 describes, spurred on by both environmentally concerned Chinese officials and the country's participation in the 1972 UN Conference on the Human Environment, China's leaders began the long, slow process of developing a formal environmental protection apparatus both at the central and local levels in the 1970s and 1980s. At the same time, they began to build a legal infrastructure for environmental protection, drafting increasingly sophisticated laws, signing onto international environmental agreements, and, beginning in the 1990s, training lawyers and judges on issues of environmental law.

Yet challenges remain. Beijing has still not sufficiently empowered its environmental protection infrastructure to meet the vast array of environmental challenges. The State Environmental Protection Administration and local environmental protection bureaus remain weak and underfunded, often incapable not only of advancing their interests in the face of more powerful development-oriented bureaucracies but also of utilizing the nascent legal infrastructure to protect the environment.

Moreover, devolution of authority to local officials, which contributed to such dramatic economic development in many regions of the country, has produced a patchwork of environmental protection. Some regions have moved aggressively to respond to environmental challenges, while others, as in the Huai valley, have been far slower to develop the necessary policies and implementation capacity.

Some of the difficulty rests in the nature of the political economy at the local level. Local government officials often have close ties with local business leaders and indeed may be part-owners in local factories. It is difficult to police polluting factories that are at least partly owned by the local government and directly contribute to the wealth of local officials. Moreover, local environmental bureaus are embedded within the bureaucratic infrastructure of the local governments; enormous pressure may be brought to bear on local environmental protection bureaus by officials concerned with maintaining high levels of economic growth no matter the environmental costs. In some respects, this devolution of authority perpetuates a form of the same traditional, personalistic system of environmental protection that characterized Imperial China and produced wide regional variation in environmental protection.

In addition, in the 1990s, as the Chinese leadership gradually withdrew from its responsibility to meet all the social welfare needs of the population, including health, education, and environmental protection, it welcomed greater public participation in environmental protection. As chapter 5 explores, China's leaders have allowed the establishment of genuine nongovernmental organizations (NGOs), encouraged aggressive media attention to environmental issues, and sanctioned independent legal activities to protect the environment, partly to compensate for the weakness of its formal environmental protection apparatus. Grassroots NGOs have sprung up in many regions of the country to address issues as varied as the fate of the Tibetan antelope, the deterioration of China's largest freshwater lakes, and mounting urban refuse. And nonprofit legal centers have emerged to wage class action warfare on behalf of farmers and others whose livelihood and health have suffered from pollution from local factories.

The media have played a crucial role, directing a harsh public spotlight on enterprises that refuse to respect environmental laws, investigating the accuracy of government accounts of environmental cleanup campaigns, and exposing large-scale abuses of the environment such as rampant logging. In this way, the Chinese leadership hopes to fill the growing gap between the people's desire to improve the environment and the government's capacity or will to do so.

The scientist Su Kiasheng and the media that challenged the government's story concerning the water quality of the Huai River reflect the potential of society to become more vocal in its demands for a cleaner environment and to bring a new level of accountability to the Chinese government's environmental policies and proclamations, and perhaps even to the entire Chinese political system.

China's leaders have also looked abroad for inspiration and assistance. Chapter 6 describes how China has eagerly gobbled up technical and financial help from international institutions, as well as other countries. China is the largest recipient of environmental aid from the World Bank, the Asian Development Bank, the Global Environmental Facility, and Japan. The international community has supported China in virtually every aspect of its environmental protection work—the development of a legal system, monitoring systems for air pollution, exploring energy alternatives to coal, and so on.

Yet, deepening integration with international actors carries with it a set of unpredictable challenges. Involvement by the international community brings not only financial assistance and technical expertise but also an array of political and social considerations that often diverge from the interests of many within the Chinese government. Infrastructure projects involving massive forced resettlement or programs involving minority populations, for example, often alarm international donors and focus unwanted international attention on China's human rights record. China's weak incentive system and enforcement apparatus for environmental protection also often thwart effective cooperation. Corruption siphons off money and goodwill from both donors and intended recipients.

Inviting such participation from both domestic and international actors outside the government is also an enormous gamble in other respects. Chapter 7 offers examples from outside China that show, once unleashed, such independent social forces may be very difficult to contain. In the former Soviet Union and Eastern Bloc countries, intelligentsia, including writers, scholars, scientists, and other professionals, as well as college students, used environmental protection as a cover for the sharing of broader grievances, and environmental NGOs served as a "school of independence for an infant civil society," as well as a training ground for liberally minded individuals to become "activist citizens" and "alternative elites." Many environmental NGOs in these countries allied with other social interest groups to focus more directly on revolutionizing the political system.

Similarly, in the less dramatic cases of China's Asia Pacific neighbors, environmental NGOs have been consistently at the forefront of political change. In advancing the cause of environmental protection, they have exposed the institutionalized corruption and lack of accountability of the entire system of governance. In countries such as Korea, the Philippines, and Thailand, during the 1970s and 1980s, environmental activists became closely linked with the democracy activists who agitated for broader political reform. In these instances, environmental issues did not serve directly as a catalyst for regime change but did permanently enlarge the political space for social action. Already, several of China's leading environmental activists leave no doubt as to their desired outcome: not only the greening of China but also the democratization of the country.

Thus, China's path to its current environmental crisis and its environmental future depend significantly on how not only the central government but also local officials, citizens, and the international community manage the environmental legacy of the past and elect to respond to the challenges and opportunities that reform presents for environmental protection today. Chapter 8 explores three scenarios outlining how the choices they make will shape China's future environment and have broader implications for the country's social, economic, and political future.

Assessing the Trends

Looking back over the period since China began its process of reform, trends in environmental pollution and degradation alone would suggest that the reform process has contributed to a significant deterioration in China's environmental situation. Yet, scratch beneath the surface and an additional, more useful set of observations concerning China's current and future environmental challenges emerge.

First, while the reform period marks a break in both the nature of the environmental challenges that China confronts and the opportunities it has to respond to them, important continuities remain from centuries-old practices and ideas concerning environmental protection. In some respects, the reforms reinforce these traditional approaches to environmental protection: maintaining a campaign mentality, relying on local officials for core aspects of environmental protection, and viewing the environment as an issue of national security.

Second, the Chinese government has made dramatic strides in developing institutions and a legal system to protect the environment, but the nature of Chinese politics, including endemic corruption and a lack of transparency, often renders the system dysfunctional.

Third, within China, there is extraordinary variability in environmental pollution and degradation statistics. As others have discovered looking at issues as varied as agriculture, automobiles, and technology, the decentralized nature of the state and decades of devolution of power to local officials virtually ensure that policy choices will be made and implemented differently from one region to the next.[62]

In a few select regions of the country, the reform process has already produced dramatic new benefits for China's environment and society. Still, in many others, it has contributed to a slide deeper into environmental crisis. Those regions in which reform has produced a positive environmental outcome, where pollution and environmental degradation are being aggressively and effectively addressed, share some common features: (1) the top local official supports or is perceived by other local officials to support environmental protection goals; (2) there is a strong level of environmental support provided by the international community; and (3) the domestic resources available to local leaders to address environmental challenges are significant. The losers, in turn, are those regions with much weaker links to the international environment and development community; whose leaders are still overwhelmingly concerned with economic development and perceive the environment as a costly luxury; and, often, whose local resources to invest in environmental protection are far less than those in other parts of the country.

Finally, as the interplay of political and economic reform exerts an impact on China's environment, it becomes clear that the environment then too comes back to influence the reform process in critical ways. Most clearly, environmental degradation and pollution constrain economic growth, contribute to large-scale migration, harm public health, and engender social unrest. Moreover, there is the potential, already evident in nascent form, for the environment to serve as a locus for broader political discontent and further political reforms, as it did in some of the former republics of the Soviet Union and in some countries in Asia and Eastern Europe.

China's Rising Star: Implications for China and the World

In the early years of the twenty-first century, China's leadership can claim much to celebrate. Fifty years of leadership by the Communist Party have brought unparalleled economic growth and well-being for the Chinese people, the expansion of Chinese influence throughout the Asia Pacific and beyond, and the return to Chinese control of Hong Kong and Macao. Only Taiwan remains to fulfill Beijing's dream of a Greater China. China's rise to major player status on the world stage was confirmed in 2001 as the country completed negotiations to join the World Trade Organization, hosted the Asia Pacific Economic Cooperation forum summit, and won the right to hold the 2008 Olympics in Beijing.

But these successes come at a heavy price. Ignored for decades, even centuries, China's environmental problems now have the potential to bring the country to its knees economically. Estimates of the costs of environmental pollution and degradation range from 8 percent to 12 percent of GDP annually. Moreover, pollution and resource scarcity have become major sources of social unrest, massive migration, and public health problems throughout the country.

How China's leadership balances its desire for continued economic growth with growing social and political pressures to improve its environmental protection has profound implications for the world beyond China. Resolution of the world's most pressing global environmental challenges, such as climate change, ozone depletion, and biodiversity loss requires the full commitment and cooperation of China, one of the world's largest contributors to these environmental problems. Regionally, China's economic development contributes to problems of transboundary air pollution, fisheries depletion, and management of international water resources such as the Mekong River. The continued stability of China—a major global trading power and pillar of East Asian security—also is drawn into question by the relentless press of sociopolitical problems, some of which have roots in the government's failure to address environmental concerns.

Beyond this lies the fundamental question of the future of China's political and economic system and its orientation toward the rest of the world. The authority and legitimacy of the Communist Party depend on how well China's leaders provide for the basic needs of their people and improve their standard of living. Fundamental to the state's capacity to meet this challenge is the environment. Access to resources for household and industrial consumption, improvements in public health, and continued economic growth are all predicated on Beijing's ability to slow or reverse the forces of environmental pollution and degradation. Thus, the environment will be the arena in which many of the crucial battles for China's future will be waged.

CHAPTER TWO

A LEGACY OF EXPLOITATION

The environmental challenges China faces today result not from decades but from centuries of abuse of the country's natural resources. Nation building, war, and economic development have all exerted unrelenting pressure on land, water, and forest resources over the country's history. As early as the seventh century, China's population also began to take a toll on the environment. Through the centuries, in turn, exploitation of the environment contributed to the cycles of war, famine, and natural disasters that plagued China and hastened the disintegration of one dynasty after the next.

Underpinning China's current environmental challenge, moreover, is a deeply rooted cultural tradition that accords little value to some of the core elements of effective environmental governance: independent scientific inquiry, a transparent political system, and accountable leadership. As historians of science in China have described, Chinese culture, rooted in Confucianism and later reinforced by Marxism-Leninism, hampered the development of modern scientific rationalism.[1] Chinese scholars' concern with preserving doctrinal orthodoxy also constrained their ability to question freely. The harsh punishments meted out to those who challenged the prevailing dogma also severely inhibited "personal responsibility, initiative, and risk-taking."[2] Furthermore, the imagery and doctrine of Confucianism have prompted exploitation rather than conservation of nature's resources.

Of course, China is not alone in its wanton environmental practices and the devastation they have wrought. As Derek Wall notes in his history of environmentalism,

> Numerous ancient societies have collapsed because of environmental degradation. The builders of Avebury and Stonehenge seem likely to have caused massive deforestation, leading to soil erosion, climatic change and probable famine. The Mayan pyramid builders may have caused their own demise in a similar fashion. Over-zealous irrigation schemes that drew salt into the soil hastened the collapse of Sumerian society and possibly that of the Indus valley.[3]

A century ago, the United States was grappling with many of the same problems that currently confront China: rapid deforestation in the midwestern states, water scarcity in the west, soil erosion and dust storms in the nation's heartland, and loss of fish and wildlife.[4] These challenges sparked a number of grand-scale public and private initiatives to conserve the land, water, and forest resources, as well as the biodiversity, of the country.[5]

But what sets China today apart from the United States of a century ago, and from other countries currently at the same level of economic development, is the scale of the environmental degradation it confronts and the magnitude of the social, political, and economic challenges this degradation has engendered. No other country confronts the gargantuan task of meeting the needs of almost a quarter of the world's population on a land mass roughly the size of that of the United States.

The Environmental Tradition

The search through history for a cultural tradition of environmentalism in China yields a rich trove of artistic and literary references to the importance and beauty of nature. Chinese painting and poetry are replete with images depicting man's reverence for the natural environment. In Chinese painting, for example, respect for nature is reflected in the "smallness" of man relative to that of mountains, rivers, and trees.[6] Artists refrained from positioning people at the center of their paintings in order to counter the notion of man at the center of the universe.[7] Throughout history, wilderness symbolized sanctuary and refuge from the political strife of the times.[8] Mountains were especially symbolic in the major philosophies and religions of China as places where one could restore one's energy as well as one's understanding of the "good and the true."[9] Even buildings were sited to ensure harmony "with the local currents of the cosmic breath."[10] The great Chinese philosopher Confucius, for example, observed, "The wise find joy in water; the benevolent find joy in mountains."[11] This evidence of a deep appreciation for nature in ancient Chinese art and literature has prompted some environmental historians to suggest that China's environmental tradition was dominated by a respect for nature.[12]

Indeed, as early as the Western Zhou dynasty (1115–1079 B.C.E.),[13] the official elite demonstrated an appreciation for the need to protect China's environment. According to "Rite of Zhou: Regional Officer," local governors were responsible for protecting rivers, mountains, forests, birds, and other animals.[14] Several centuries later, Guan Zhong, the prime minister of the Qi State (?–645 B.C.E.) in the Spring and Autumn period of the Eastern Zhou dynasty, espoused values and practices that respected the relationship between humans, development, and environment. He advised the people to utilize the forest and the fisheries, but only to a "reasonable extent."[15] He also cautioned people "not to raise too many cattle on the grassland, lest it fail to recover from over exploitation; and not to plant crops too close together, otherwise the fertility of the soil would be insufficient."[16]

In contrast, Chinese folktales more often portrayed nature as a force to be overcome and utilized for human purposes. In one myth, for example, the great archer Hou Yi encounters ten suns burning so brightly that they threaten life on earth. He bravely shoots down all but one of the suns, thereby preserving earth and earning the elixir of life from the Queen Mother of the West.[17] Another story recalls the triumph of the Great Emperor Yu, who, seven thousand years ago, held back the floods of the Yellow River by dredging the waterways. His success gave rise to the belief that if people adjust their actions to circumstances, they can conquer nature's scourges, just as "Dayu [the Great Emperor Yu] regulating rivers"[18] did.

China's environmental tradition was also influenced heavily by the leading schools of early Chinese thought: Confucianism, Taoism, and Legalism. Buddhism, while not Chinese in origin, was also influential. Each school embraced earlier notions of man's relationship to nature as expressed in art, literature, and practice, and each brought to these ideals and practice its own perspective. In turn, these philosophies influenced Chinese authorities, elite and popular attitudes, and the overall ordering of Chinese society in ways both distinctive and important for the natural environment. Yet, as China historian Charles Hucker notes, while these schools of thought had "different points of view and emphases . . . , they were not mutually exclusive. Confucianism focuses on man in his social and political relationships, Taoism on man's status in the larger cosmic sphere, and Legalism on state administration."[19]

The Philosophical Underpinnings

To a significant extent, each of the major Chinese approaches to organizing social relations within the broader cosmos was rooted in earlier practices of the Western Zhou dynasty, which established principles regarding the human relationship to the cosmos that the great philosophers and statesmen of later dynasties would build on and modify. The Zhou believed that the cosmos was ruled by an all-powerful heaven (*tian*), which in turn entrusted one man, the King of Zhou or the son of heaven (*tian zi*), with responsibility for managing "all under Heaven." The Chinese world was to be united under this ruler, whose mandate was to govern with the counsel of wise advisers and bring peace and order to his domain.[20]

During the Spring and Autumn period of the Eastern Zhou (722–481 B.C.E.), the emperor was expected to respect the environment: "People who are of ruling quality but are not able to respectfully preserve the forests, rivers, and marshes are not appropriate to become rulers." And "In spring, if the government does not prohibit (cutting) then hundreds will not grow. In summer if the government does not prohibit (cutting) then the crops will not succeed."[21]

Confucianism

Confucius' vision of social organization emphasized the importance of morality, propriety, and social harmony.[22] Confucius imputed morality not only to people but also to the cosmos. He made the linkage thus: The universe and mankind are governed by an impersonal but willful heaven (*tian*); heaven wills that men be happy and orderly in accord with the cosmic harmony (*Tao*);[23] and an "ethical and virtuous life is the appropriate human contribution to the cosmic harmony."[24] The cornerstone of his philosophy was the "five fundamental relationships": parent to child, husband to wife, elder sibling to junior sibling, elder friend to junior friend, and ruler to subject. As long as the first subject in each pair provided the proper example for the second, harmony and freedom from conflict were assured.[25] A leader thus established order by being a moral exemplar, and government as a whole operated by example rather than by law.[26]

Nature occupied its own space in this filial ordering. The Confucian scholars Mencius and Zi Si (Tzu-Ssu), while developing the idea that humankind and nature were inseparable (*tian ren he yi*), also suggested that *tian*, encompassing the sky, God, and nature, was superior to humans, and that only raising the level of human morality could enable us to understand the objective laws of nature.[27] Xunzi (Hsün-tzu), the Chinese philosopher of the Warring States period (403–221 B.C.E.), however, conceived of humans and nature as occupying two different worlds, and he argued that nature would not change its objective laws because of human will: "*Tian* would not stop the coming of winter because of my hatred to coldness. *Di* [the earth] would not stop being broad because of my hatred to broadness."[28] At the same time, he articulated a belief in the human ability to meet nature's challenges and control it for human needs.[29] He developed the notions of "*Zhi tianming er yungzhi*" (master nature for use) and *"Ying shi er shi zhi"* (follow the law of the seasons in agriculture).[30] He believed that by appreciating nature's laws, man could overcome nature's superiority and use it for his benefit.[31]

During the Qin (221–207 B.C.E.) and Han (202 B.C.E.–220 C.E.) dynasties, philosophers continued to refine their understanding of the interaction among the divine, natural, and human worlds.[32] Qin scholars and philosophers, under the tutelage of Lu Buwei (Lü Puwei), prime minister of the First Emperor of the Qin, produced the comprehensive work *The Spring and Autumn of Mr. Lü* (*Lü-shih ch'un-ch'iu*),[33] which established the emperor (*tianzi*)[34] as the intermediary among the three worlds.

Nature possessed a godlike ability to wreak havoc or bring bounty as punishment or reward for humans' behavior. As the spiritual and temporal leader of all mankind, the emperor was responsible for ensuring that every phase of human activity proceeded properly and remained in harmony with the workings of the other two orders.[35] Emperors took this responsibility seriously. According to one tale, during the Shang dynasty (1766?–1122? B.C.E.), for example, "Emperor Shangtang decided to sacrifice himself after his country had suffered drought for seven years. At the moment he lit the firewood to burn himself, *tian* was moved and rained to put out the fire. The drought was then relieved."[36]

Scholars did not merely content themselves with abstract philosophical ideals, however. Mencius, for example, counseled King Hui of Liang: "Plant your crops according to the changes of season, you will have more food than you want; limit the size of the holes in your net, you will have more fish than you want; bide your time to fell the trees, you will have more fuelwood than you want."[37]

Thus, in the Confucian conception, the world was governed by a triad of heaven, which was the "creative force with an effective will"; earth which was to "bring forth what Heaven generated, and to support and nourish," and mankind in the middle.[38] Earth had the critical role of providing for man's benefit. In practical terms, "it was up to man to invest toil and ingenuity in realizing Earth's potential, thereby bringing forth the means of mankind's sustenance, and creating an admirable material civilization fit to awe barbarians. The state justified its existence by leading society in this endeavor."[39]

Historian Rhoads Murphey elaborates:

> In this classic agrarian society, the Chinese saw the land and the agricultural system which made it yield as the *summum bonum*, both materially and symbolically, the source of all value and of all virtue. The bulk of imperial administration was devoted to its care, directly or indirectly, and agriculture was the support of state and society probably to a greater degree than in any other civilized tradition since ancient Egypt. The emperor's most important ceremonial function was the yearly rites at the Temple of Heaven in the capital, where he ploughed a ritual furrow and interceded with heaven for good harvests . . . most of the gentry-literati . . . supported a deep respect, even a reverence, for a natural order conceived as grander than man and more to be admired. Admittedly this frequently had didactic overtones: a man should respect and admire nature as he must respect and admire his human superiors, specifically his parents and the official hierarchy which was symbolically viewed in surrogate parental terms. The proper attitude in both cases was filial. To question or attack nature was to contradict a broader natural order, sometimes labelled "heaven" (t'ien), and hence was potentially disturbing to the profoundly hierarchical social order of traditional China.[40]

Taoism

In contrast to the highly structured, morality-based system of Confucius and his disciples, the Taoist credo was "Let Nature take its course! Be yourself! Relax and enjoy life!"[41] As Hucker comments,

> Taoism is a lyrical, mystical, but by no means irrational advocacy of individualism, quietism, and spontaneity in union with Nature (tao). . . . Nature is conceived as all-encompassing, as an impersonal, purposeless cosmos in which everything has its natural

place and function; it is what it is for no reason other than that it is what it is, and can only be distorted and misunderstood when it is defined, labeled, or evaluated by standards such as good and bad that do not exist in nature.[42]

Taoism embraced a life of simplicity; man should live as close to nature as possible and do nothing with regard to organizing or managing affairs of state.[43] As Laozi (Lao-tzu) argued in *Dao De Jing* (*Tao-te Ching*): If it was a man's "lot" in the "natural scheme of things" to be a ruler, then he should keep the people "rustic and simple"—"fill their bellies and empty their minds, strengthen their muscles and weaken their wills."[44] As Ray Huang points out, the taoists' "acceptance of cosmic unity and readiness to return to primitive simplicity were reinforced by their resistance to the curtailment of freedom, either through enticement or through cohesion [*sic*]. Taoism, therefore, gave comfort to pantheism, romanticism, and not the least anarchism. Those sentiments, however, provided no immediate cure for the current political turmoil except to turn wise men into recluses."[45] Such a philosophy might protect the environment, but it would also delay economic development.[46]

Neo-Confucianism

The writings of later neo-Confucian scholars such as Zhang Zai (Chang Tsai; 1020–1077 C.E.) and Wang Fuzhi (Wang Fu-chih) combined the sense of order and morality prescribed by Confucian thinkers with a strongly Taoist "oneness" with nature: "Heaven is my father and Earth is my mother, and even such a small creature as I find an intimate place in their midst. Therefore that which fills the universe I regard as my body and that which directs the universe I consider as my nature. All people are my brothers and sisters, and all things are my companions."[47] For the neo-Confucians, human beings were "organically connected with rocks, trees, and animals and formed one body with the universe."[48] As Robert Weller and Peter Bol articulate,

Neo-Confucianism continued to base ideas of personal and political morality on an understanding of "Heaven-and-Earth" as an integrated, coherent organism, largely consistent with the lines initially established by cosmic resonance theory. At the same time . . . the natural world [was] treated in practice as a metaphor for the integration and coherence that humans should try to establish in social life; it generally did not lead scholars to the investigation of the coherence of the natural order or to a biocentric view of the world.[49]

Although modern neo-Confucians have tried to demonstrate an environmentalist spirit in their intellectual heritage, their predecessors were not environmentalists as we understand the term today (i.e., people committed to protecting the environment from pollution or degradation).[50] Rather, they were interested in exploring the place of human beings within the natural world and in establishing that people, as the "most advanced creatures," were "biologically endowed with principles for structuring perception, thought, and action that would, if realized, result in an integrated social world."[51]

Legalism

Legalism introduced yet a third, distinctive approach to governing human relations, one that developed over time, beginning as early as the seventh century B.C.E. The "final legalist synthesis" was achieved in practice by Li Si (d. 208 B.C.E.), chief counselor under the First Emperor of the Qin.[52]

Legalist thought was in some ways centrally rooted in the state of China's natural environment at the time. The Legalists argued that as China's population grew larger and food and other goods became scarce, strict controls and rules were needed. They believed that only the law could ensure that both the ruler and the ruled would do what was in the interest of the state as a whole. Laws were to be promulgated clearly and in great detail, with a system of rewards and punishments to encourage appropriate behavior.[53] Influenced by the Legalists, the Qin dynasty instituted a system of merit-based appointments for government service.

When the Qin collapsed, however, the Legalists fell into disfavor, and their system of governance was adapted to reincorporate many Confucian ideals. For example, once-absolute laws became far more relativistic and conditional. The magistrate was expected to consider every aspect of a case, including the circumstances, motivations, and consequences of the crime; "Quibbles about the letter of the law were not important."[54] Thus, the punishment for a local dyeing factory for polluting the water of nearby peasants might take into account how many people the dyeing plant employed and the impact of a severe fine on their livelihood.

Buddhism

Buddhism entered China via India by the second century C.E. and informed the Chinese approach to nature in a very different way. It encouraged reverence of nature by introducing the notion of equity between humans and other creatures.[55] In reincarnation, humans could return to the earth as an animal or insect; Buddhists, therefore, did not kill animals, and Buddhist monks were vegetarians. Buddhist monks also protected the environment in the mountains in which they typically lived. According to one Chinese scholar, the fact that "one-third of the most beautiful and famous scenic areas in China [today] are Buddhist temples surrounded by ancient trees" reflects the harmony between man and nature espoused by Buddhism.[56]

A Ravaged Land

Confucianism, Taoism, Legalism, and Buddhism share a healthy respect for the importance and power of nature to shape man's conditions and prospects for a fruitful and prosperous life. Yet it is the Confucian belief in man's ability to shape nature to fulfill his needs that is most evident throughout Chinese history. The efforts of early environmental thinkers and officials were overwhelmed by the imperatives of war, economic development, and population growth. Thus, the continual cycles of social transformations, including war, population growth, economic development, and eco-environmental change[57] resulted in astonishing levels of deforestation, desertification, soil erosion, and flooding.

Indeed, environmental historian Mark Elvin questions whether China ever achieved a balance between humans and the environment: "The restraint preached by the environmental archaic wisdom found in certain Chinese classical texts is both familiar and in all likelihood commonly misunderstood: it was probably not a symptom of any ancient harmony but, rather, of a rational reaction to an incipient but already visible ecological crisis."[58] As early as the Xia,[59] Shang (1766?–1122? B.C.E), and Zhou dynasties, overexploitation of the land was evident: "As people completely relied on the natural productivity of the land, farmland was intensively cultivated to obtain maximum production."[60] The problem of "migratory farming" was especially severe as early Chinese peoples continuously relocated to find better land.[61]

Elvin describes a China in which the quest for power by both the state and the individual transformed the natural environment for the purpose of warfare. This transformation involved, for example, rampant deforestation for fuel for burning coal and intensive exploitation of the mountains and grasslands for ores to produce spears, cannons, and armor.[62] As he pointedly states,

> The Chinese search for power—both by the state and individual—was based, until very recent times, primarily upon an almost unending transformation of the lowland landscape to adapt it to an intensive agriculture, making use of hydraulic technology in the form of flood control, drainage, irrigation and seawalls to stabilize production in the face of irregularities in the weather, both seasonal and between years. . . . It has been the pay-off in power accruing to the "exploitation" of the environment at a rate overstressing its natural resilience and exceeding its capacity for self-renewal within a humanly relevant time-frame that has . . . made the process hard to restrain by conscious action, even when there has been a fairly widespread appreciation of its damaging effects.[63]

This was particularly evident during the 248 years of the Warring States period in which there were 590 recorded wars.[64] And, the

Spring and Autumn Annals offer numerous references to war resulting from opposing armies raiding each other's crops and cutting off food supplies in time of famine.[65] The Annals also reveal that Qi State prime minister Guan Zhong convinced the leaders of other principalities never to "execute water works detrimental to the interests of the other states and never to impose a grain embargo in time of famine." But while Mencius, some 350 years later, cited the resulting Guiqiu Convention as a model for rulers of his time, he noted that such wise counsel was typically ignored.[66]

The ever-expanding search for fertile land also drew states into increasing competition and conflict. As Huang recounts, "On one occasion, perhaps in 320 B.C.E., the king of Wei, whose territory embraced both banks of the Yellow River, told the Second Sage [Mencius] that in times of serious crop failure he had to relocate the populace across the river in large numbers. By this time, the state of Lu had expanded its territory from its original authorization by five times, and neighboring Qi, by ten times."[67]

Even the consolidation of power after war often proved devastating to the environment. The first emperor of the Qin united six states and formed the first centralized state in Chinese history, an accomplishment that historians of China laud to this day. The emperor and his officials also constructed major irrigation works and canals to promote economic development and assist in the centralization of the state.[68] Yet the reality of state building placed an enormous burden on both the natural environment and the peoples of the time. Geping Qu and Jinchang Li detail:

> Unification by military force killed more than a million people and lay waste to 13 cities. Deaths related to hunger and displacement outnumbered those related to battle. After unification, large-scale construction detracted from other productive activities, leaving people with little respite. About 400,000 people helped build the Great Wall. Another 500,000 guarded mountains and suppressed riots. And 700,000 built the E'Fang Palace and Qin Shihuang's tomb at Lishan Mountain (site of the Terracotta Warriors). Diggers became sacrificial objects themselves. As a result, there were an insufficient number of men for farming and women for weaving.

Although unification of the Qin was a great contribution to history, its rulers extorted taxes and forced military service on the people. The population of the Qin fell to below 20 million. Construction efforts during this period caused large-scale environmental degradation to forests and other natural resources. The poet Du Mu of the Tang dynasty approved of unification and the glory of E'fang, but lamented deforestation: "Six states conquered, the world was united . . . with Shushan Mountain bare, E'fang Palace rose to glory."[69]

Thus, if war was one source of environmental degradation, the aftermath of war and the process of economic development was another.

The Han dynasty (202 B.C.E.–220 C.E.) illustrates how prosperity in China contributed to environmental degradation.[70] Population growth increased the demand for food. As arable land became insufficient, officials pushed the populace further afield to cultivate wasteland; more than 20 million acres of land was reclaimed for farming.[71] Local gazetteers during this time recorded the gradual deterioration and exploitation of the environment in individual regions. During the Han, for example, the people of Jiaxing, in the lower Chang (Yangtze River) delta, progressed from a situation of environmental affluence, where "food, including fruits and shellfish, was so abundant that the poor are said to have lived from day to day without keeping reserves," to a point in later imperial times where "economic, environmental and social stress is evident everywhere."[72]

As China's economy flourished during the Tang (618–906 C.E.), Song (960–1279 C.E.), and Yuan (1276–1367 C.E.) dynasties, settlers pushed southward, causing deforestation and soil erosion.[73] Still the needs of the people failed to be met. Qu and Li detail the unrelenting pressures of war followed by rapid development during these periods:

> Civilization was subsequently wrecked by natural and anthropogenic calamities, frequent wars, and general turmoil. The prosperous Yellow River Valley became desolate; water-saving facilities and farmland lay to waste; masses emigrated south or starved. . . . [74]

[Even] in the heyday of the Tang dynasty, food production fell short of the needs of 53 million people. To meet the shortfall, 6.2 million ha [15.3 million acres] of land were reclaimed for agriculture, severely damaging many fragile ecosystems. . . . Soil-erosion, river expansion, and flooding were other unintended costs. The Yellow River overflowed its banks more than 50 times in 100 years of Song dynasty rule.[75]

During the Ming and Qing dynasties, deforestation, flooding, soil erosion, and desertification all increased as the economy continued to develop:

New approaches to agricultural production wreaked havoc on the environment; . . . entire forests were harvested for cultivation; reclamation additionally occurred on the frontier, prairies, mountains, and islands—all unique areas that buffered densely populated regions. . . . Employing various harvesting methods, the peasant farmers would leave nutrient-rich soils infertile after several years and deforest new areas. Centuries old forests north of the Qingling Mountains were left treeless. . . . An additional sign of environmental degradation includes unprecedented soil erosion. Sand content in the sedimentary layers of the Yellow River reflects soil erosion and frequent overflows. During the Ming dynasty, the river overflowed 127 times, approximately once every 2 years. Later, during the Qing dynasty, the river continued to flow over its banks, flooding more than 180 times in a 200 year period. . . . Cultivated areas and cities of the Han and Tang dynasties in the North and Northwest were literally submerged by sands during the Ming and Qing reigns. . . . The Badan Jilin, the Ulanbuhe, and the Maowusu deserts continuously expanded and destroyed vegetation. Desertified areas of Kergin in Northeast China were more or less created by land reclamation during the Qing.[76]

Beginning in the mid-seventeenth century at the onset of the Qing dynasty, grounds and parks were set aside for sightseeing and recreation, a practice that helped to protect animal and plant species that otherwise would have been endangered.[77] Yet the net effect of economic development during the Qing was an environmental disaster. There was a dramatic population increase, excessive cultivation of land, overherding, overfishing, and overcutting, all of which caused serious damage to forestry, agriculture, and fishery resources.

By the late 1800s, as Elvin details, China was a country wracked by "the exhaustion of resources, especially wood and mineral ores, shortages of water, erosion and lands ruined by salinization resulting from inappropriate development."[78] There was a "desperate anxiety about the supply of wood, and its quantity and quality . . . [as] the rate of depletion, driven by perceived economic need, exceeded the rate at which stocks renew[ed] themselves either solely naturally or with the help of an affordable silviculture."[79]

Natural disasters compounded these challenges. In 1876, thirteen million people died after a three-year drought. Ten years later, two million more people died when the Yellow River overflowed its banks.[80]

The Burgeoning Populace

The desire of successive emperors to increase the size of the population was a third contributing factor to the pressures exerted by man on the natural environment. Nearly all Chinese leaders regarded a large population as a necessity for a strong state, contributing both to greater tax revenues and a larger army. Shang Yang, the statesman that guided the development of the Qin state, argued that a large population and a strong army were the greatest assets of a ruler. Liu Yan of the Tang dynasty argued that if there were more people, there would be more tax payments. Similarly, Ye Shi of the Southern Song dynasty opined that "the most essential thing for a state is to have more people, because more people would till more land and pay more taxes and the army would have more recruits." And Qiu Jun of the Ming dynasty stated that "a peaceful life meant proliferation of people and if the population increased, the fundamental power of the state would be consolidated and the national strength would grow, thus bringing about great order and safety and stability to the throne."[81]

Not surprisingly, China has repeatedly confronted population issues. Elvin dates the problem to the Tang (618–906 C.E.) and Song (960–1279 C.E.) dynasties,[82] and Helen Dunstan as far back as at least the Sui dynasty (589–618 C.E.).[83] As Dunstan notes, "Confucian humanism was all in favor of the flourishing and propagation of the human species. For a region to be 'prosperous and teeming' was a positive phenomenon."[84] Mencius, for example, believed that "having no children" was one of the "three most impious acts."[85]

A few statesmen and scholars challenged the conventional wisdom concerning the value of a large population. Guan Zhong, for example, argued that population increases had to be regulated by two key ratios: land and the composition of the population. In the first instance, too many or too few people per unit of land would cause inefficiencies and potentially food shortages. "A good ruler," he argued, "knows well how many people he reigns over; only sufficient arable land will make his people content."[86] In addition, balance between urban and rural populations and military and civilian populations was crucial. If there were too large an urban population, "there would not be enough farmland to support the population." Similarly, if there were too large a military, it would "leave base land uncultivated and also bring the nation to the brink of poverty."[87]

The Ming dynasty epitomized the impact of rapid population growth on the environment. During this period, the population expanded from an estimated 66 million in 1403 to approximately 100 million by the final years of the Ming in the mid-1600s. There were reports of widespread deforestation, massive land reclamation projects, soil erosion, and flooding as peasants left "nutrient-rich soils infertile" and moved to deforest other areas.[88] As a result, "floating" manual laborers and entrepreneurs migrated to the urban coastal areas in search of work.[89]

By the mid-1700s, Chinese officials formally acknowledged the difficulties of adequately providing for a large population by explicitly recognizing the link, for example, between the growing population and the increasing price of grain.[90] Yet it was not until a century later that Qing scholar Wang Shiduo proposed a set of measures to reverse the population trend. He advanced such draconian measures as the death penalty for men who married under the age of twenty-five and women who married under the age of twenty, a tax incentive for the infanticide of second daughters, and compulsory abortion.[91]

The Search for a Response

Throughout these centuries, China's system of governance was clearly inadequate to the demands of environmental protection. Most responsibility for the environment lay with the provincial administrators, who made it an exceedingly low priority. For a Qing administrator, for instance, the two most important roles were the administration of justice and collection of taxes. He was also responsible for overseeing population registration, security, postal stations, public works (such as reservoirs and dams for local farm irrigation and roads), granaries, famine relief, poorhouses, and ceremonial observances, including sacrifices to the gods of earth and grain, wind, clouds, thunder, rain, mountains, streams, and agriculture. Finally, local magistrates were also charged with encouraging the people to farm and cultivate mulberry trees, dredge rivers, and build dams.[92] Yet few officials took the latter set of responsibilities seriously. One imperial edict complained that "it was difficult to find officials who would concern themselves with such things as farming and irrigation, which in ancient times had been encouraged by good officials."[93] As Dunstan comments, "Discussions of how to collect the land tax both successfully and with maximum sensitivity to peasant interests can be detailed, logical, perceptive and exhaustive. Environmental thinking was commonly attenuated by contrast."[94]

Nonetheless, some provincial officials took a serious interest in environmental protection. In Henan province during the late 1600s, for example, Yu Sen, the province's assistant surveillance commissioner with special responsibility for watercourses proposed a massive reforestation program to address some of the devastation wreaked by the late Ming rebellions. While some aspects of the plan were not necessarily practical, his writings demonstrated a clear understanding of how willow trees could contribute to control soil erosion: "The soil of He'nan is not firm . . . and river banks are liable to become eroded. If one plants rows of willow trees, the roots will intertwine [so that] the protective dike is solid. At what point will it be possible for [it] to be washed away?"[95]

Yu also evidenced a broader philosophical approach to man's relationship to nature:

> The functioning of the five phases [wood, fire, soil, metal, and water] is such that failure to conquer means failure to generate. If now trees are scarce, wood will not conquer soil, and the nature of soil will be light and easily blown away, while the human character will become crude and fierce. If trees are plentiful, the soil will not fly up, and men will revert to refinement and good order.[96]

Some officials also comprehended the correlation between development and environmental degradation and even developed overall plans for managing the ecosystem of a region. Chen Hongmu, Yunnan provincial administration commissioner in 1737, concerned himself with the full range of issues: reforestation, crop diversification, land ownership, and how to "brew" animal manure for fertilizer.[97] Chen's reports on reforestation indicate his clear understanding of how man had changed the landscape:

> In mountainous Yunnan there were (originally) endless bamboo forests covering the serried hills, while trees and grasses grew luxuriantly and ample firewood was there for the taking. However because the salt-works require wood, and mines and mints require charcoal, there has been daily cutting, resulting in denuded hills almost as far as the eye can see. Firewood and charcoal have gradually become dear and hard to come by; even the roots of grass and bark of trees are almost rarities. . . . Yunnan, being mountainous in all directions, has land on which grain crops cannot be grown, but it has absolutely no place in which trees cannot be planted. Besides, (since the province) was originally wooded with tall forests and dense bamboo thickets which have only now been felled almost to the last tree, reforestation should indeed be altogether feasible.[98]

At the same time, Shaanxi provincial governor Bi Yuan set forth measures to balance the burgeoning population with the increasingly scarce natural resources. He laid out a master plan for "managed settlement," which included resettling people, expanding irrigation works, and organizing a stock-rearing industry throughout the province, including camels, horses, oxen, and sheep.[99] During 1778–1779, he resettled over a hundred thousand people in the southern highlands.

The local people, too, were not oblivious to the harmful effects of poorly planned economic development on their livelihood and well-being. The damage to crops from local industry was a common concern during the Qing, and there are accounts of rural residents actively protesting environmental pollution. Dunstan recounts one case in 1737, in which 108 local residents submitted a petition to the local officials protesting that the dyeing workshops in the Huqiu hill district near Suzhou were polluting the local water supply, harming crops, spoiling drinking water, and damaging the scenic, religious spot. The authorities responded by prohibiting dyeing in the area.[100] In Yunnan, one citizen wrote a letter detailing the ways in which mining was harming the local crops by polluting the water and air and denuding the hills.[101]

Despite these scattered efforts, by the second half of the nineteenth century, the Chinese political system confronted a series of very serious challenges linked to the management of natural resources:

> Systems of water transport and control began to break down, partly through neglect but also ironically through over-expansion. Excessive dike building in central Hunan during the eighteenth century to create more farmland for the growing population restricted waterways and blocked drainage channels, leading to a protracted flooding in this rich agricultural area. Along the Yangtze, conflicting interests of officials and local elites inhibited dike repair. Water control elsewhere suffered from similar conflicts of interest and decay of old facilities. [Despite success in the earlier Qing period in raising] agricultural productivity [by introducing] improved seed strains and better farming methods, [during the later Qing and

early republican period], agricultural production per unit of land began to stagnate and fall behind population growth.[102]

These problems were compounded by the large population in some regions, a failure to invest in public works projects, corruption, riots, protests against taxes, rebellions by minority people on the borders, and pressure from the West.[103]

The 1911 Revolution, which overthrew the Qing dynasty and the Confucian-based order, did little to relieve the deteriorating social, economic, and environmental circumstances of most Chinese. Approximately 80 percent were employed in agriculture, and most lived in terrible poverty. There were famines, mass migrations between provinces, widespread rural unemployment and poverty, and forced sale of land. Farmers fled the countryside to beg on city streets, and parents were forced to sell their children.[104] Customs officials of the time provided some insight into the rapid disintegration of the state, noting "the breakdown of law and order, the disruption of transportation and marketing, and widespread, uncontrolled flooding and drought throughout the country."[105]

The Nationalists did not entirely neglect the traditional imperial duty of economic development, and many urban centers realized economic growth rates of 7 percent to 8 percent under their leadership. The Nationalist leaders undertook flood control and irrigation projects as well as agricultural research to alleviate some of the environmental and agricultural problems faced by the peasants. Yet most failed to prosper thanks to "bureaucratic boondoggling" that resulted in "little positive achievement."[106] The peasants were further overwhelmed by demands from the tax collectors and other government officials for labor, supplies, and land.

Moreover, throughout the republican period (1911–1949 C.E.), the demands of continuous warfare either for the consolidation of power domestically or against an external foe, Japan, further contributed to the sustained agricultural crisis, environmental degradation, and social dislocation. By 1930, China's death rate was the highest in the world.[107]

Devastation under Mao

In 1949, the ascension to power of the rural-based Chinese Communist Party appeared to bring some initial relief to the economy and the environment. During the first five to seven years of Communist rule, the Chinese leadership repeated the practices of imperial leaders and began rebuilding the economy after a devastating war. Both agriculture and industry developed rapidly and, at the same time, some water and soil conservation guidelines were implemented. The Communists also launched major public works projects in water conservancy, afforestation, combating natural disasters, and other fields. Not all of these programs were effective. Liang Xi, the first minister of forestry, for example, stated in 1956, "As everyone knows, our afforestation statistics are not really based on actual measurements, but are fixed on the basis of mere eyeballing or guesswork. Consequently, there are mistakes, overestimates, and even totally unfounded reports."[108] Factories, especially power plants, were often built along rivers and without treatment facilities for waste water, waste gas, and industrial residue. Instead, the factories merely used the rivers for drainage.

Mao's vision for China as a great power[109] soon brought about a renewed cycle of population growth, accelerated indiscriminate mobilization of resources in preparation for war, and grand schemes for economic development, which, in turn, contributed to severe environmental degradation and social turmoil. Moreover, while Mao praised the Legalist tradition of the Qin, articulating a strong belief in the importance of laws and regulations as "instruments for procuring happiness,"[110] his early writings suggested that people might often need to be "persuaded" of the utility of such laws. In his earliest recorded essay, "Essay on How Shang Yang Established Confidence by the Moving of the Pole," written when he was nineteen, Mao noted that the popular resistance to the brutal enforcement of Shang Yang's laws could be explained by "the stupidity of the people of our country," and their reluctance to undertake anything new: "at the beginning of anything out of the ordinary, the mass of the people always dislike it."[111] Rather than translating into an enduring legal system, Mao's understanding of Shang Yang's experience set the stage for a continuous cycle of campaigns and mass mobilization efforts, several of which had devastating consequences for the environment. Ultimately, few strides were made in advancing environmental protection regulations during the decades of Mao's rule.

Mao soon forswore even the pretense of respect for nature and for man's position in the cosmic hierarchy that had at least in theory guided the rulers who preceded him. His views suggested an exaggeration of Xunzi's approach; namely, that by understanding the laws of nature man could overcome nature. As historian Judith Shapiro notes,

> The Mao-era effort to conquer nature can thus be understood as an extreme form of a philosophical and behavioral tendency that has its roots in traditional Confucian culture. . . . State sponsored resettlements and waterworks projects, extensive and excessive construction of dikes for land reclamation, political campaigns to change agricultural practices, and environmentally destructive land conversions in response to population shifts can be found in imperial times.[112]

As early as 1940, at a speech at the inaugural meeting of the Natural Science Research Society of the Border Region, Mao stated, "For the purpose of attaining freedom in the world of nature, man must use natural science to understand, conquer and change nature and thus attain freedom from nature."[113] Mao's later writings and speeches focused overwhelmingly on the need to conquer or harness nature for man's needs. As Murphey has described Mao's conception of nature, "Nature is explicitly seen as an enemy, against which man must fight an unending war, with more conviction and fervour and with a brighter vision of the ultimate results than even the Darwinian-Spencerian West held."[114]

Headlines in newspapers and magazines of the period offered evocative imagery in support of Mao's approach; for example, "The Desert Surrenders" and "Chairman Mao's Thoughts Are Our Guide to Scoring Victories in the Struggle Against Nature."[115] Mao himself penned an article in 1945 entitled, "How the Foolish Old Man Removed the Mountains," which retold the story of the legendary Yu Gong, who, in the face of scorn and mockery by his neighbors, was able to level two big mountains using only axes and the help of his two sons. In the retelling, however, Mao eliminated the conclusion, which acknowledged Yu Gong's supernatural aid from heaven, in order to avoid a contradiction with his message concerning the "boundless power of the people liberated and mobilized" to be able to move mountains.[116]

Mao's approach to the natural environment emphasized liberating and mobilizing the people to confront imminent threats from both environmental and human sources. He called on the people to prepare against war and natural disasters,[117] by digging tunnels deep, storing grain everywhere, and not attempting to dominate other nations.[118] Murphey bluntly states that this "assault on nature" was "a convenient rallying cry for an economically and technically underdeveloped society striking out, sometimes . . . blindly . . . ; the war against nature can be dramatized much more readily than the more prosaic processes of saving, accumulation, and investment."[119]

Mao's vision for China's future greatness had as its centerpiece a large population. In an August 5, 1949, white paper *The Relations of the United States and China* prepared for U.S. president Harry Truman, Dean Acheson argued, "The population of China during the 18th and 19th centuries doubled, thereby creating an unbearable pressure upon the land. The first problem which every Chinese government has had to face is of feeding this population. So far none has succeeded."[120] Mao's response to Acheson in his article "The Bankruptcy of the Idealist Concept of History" was to reiterate the utility of China's large population:

It is a very good thing that China has a big population. Even if China's population multiplies many times, she is fully capable of finding a solution; the solution is production. The absurd argument of Western bourgeois economists like [Thomas] Malthus that increases in food cannot keep pace with increases in population was not only thoroughly refuted in theory by Marxists long ago, but has also been completely exploded by the realities in the Soviet Union and the Liberated Areas of China after their revolutions.[121]

At times Mao seemed to vacillate on whether there could be difficulty in meeting the needs of too large a population. At a 1958 meeting in Chengdu, for example, Mao said, "It would not do to have too large a population and to have too [little] land."[122] One of the hallmarks of Mao's rule, however, was the dramatic increase in population from 540 million in 1950 to 930 million in 1976, an increase for which China is still paying the price, having now attained a population of 1.295 billion.[123]

Mao also mirrored his Qing predecessors in failing to heed the wise counsel of an adviser on limiting the size of the population, in this case Ma Yinchu, a distinguished economist and former president of Beijing University. In contrast to Mao's contention that "with many people, strength is great,"[124] Ma believed that if China's population continued to increase unchecked, it would become a major obstacle to the development of productivity. With population control, however, he argued "it would be not very difficult to make the lofty aspirations of building socialism a reality."[125] Ma proposed several measures, including family planning in the country's economic plans, late marriage, contraception, and a regular population census.[126]

Despite initial support from others within the scholarly community and the Chinese leadership, Ma, like thousands of other intellectuals, fell victim to the anti-rightist campaign during the late 1950s, losing his position and his livelihood. It wasn't until 1979, at age ninety-eight, that Ma was eventually rehabilitated by Deng Xiaoping: He was made an honorary president of Beijing University, and all charges against him were dropped—an acknowledgment of the wrongs perpetrated against him during Mao's tenure.[127] Such redemption, however, came twenty years too late to prevent the years of devastating consequences of Mao's population policy.

While Mao was to reverse himself many times on the issue of whether too large a population was problematic for China, it was not until 1974—when he perceived a reduction in the external threat to China from the United States and Japan and felt increasing pressure to feed and employ his people—that Mao called publicly for family planning population control.[128]

Campaigns of Destruction

Great Leap Forward, 1958–61

In 1958, Mao's belief in the ability of man to conquer nature and his desire to achieve great power status for China came together in the launching of the Great Leap Forward, a mass-mobilization campaign designed to catapult China into Communism and surpass the industrial achievements of Great Britain and the United States. The modest efforts at environmental protection of the early Communist years were quickly undermined as the country engaged in huge land reclamation projects for the purpose of planting grain, thereby laying waste to forests, wetlands, lakes, and rivers.

Qu Geping, who became the first director of China's National Environmental Protection Agency some two decades later, describes this period as one in which

> conceit began to emerge in guiding principles, and human will and capacity were exaggerated. . . . A 'leftist' tendency characterized by unrealistic targets, arbitrary guidance, exaggeration, and egalitarianism was rampant throughout the country causing great losses to the national economy and people's livelihood, as well as problems and damages to the environment and ecosystem. . . . During the Great Leap Forward, factories were built haphazardly without any consideration for environmental protection. . . . Biological resources, forests in particular, were seriously damaged, causing several losses to the ecosystem. . . . There was extensive destruction of the natural environment of our country.[129]

In hopes of raising grain yields to previously unattained heights, a wide range of misguided directives were issued. The Chinese people were exhorted to eliminate the four pests (rats, sparrows, flies, and mosquitoes); to afforest everywhere—"around every house and every village, by roadsides and watersides as well as on waste land and barren hills";[130] and to introduce new methods of raising grain yields, including close planting of seeds, deep plowing, extensive use of inappropriate fertilizers, tractors, and farm implements, and planting less (but producing more). As chronicler of the period Jasper Becker describes, many of the agricultural innovations were based on the specious science of the Soviet expert Trofim Lysenko. Some of the mistakes were mind-boggling. In Qinghai, for example, prison inmates tried to make iron-hard soil suitable for planting by "digging little holes and filling them with straw and grass which were set on fire." In Guizhou, peasants dug trenches for sowing seeds so deep (up to thirteen feet) that they had to tie ropes around their waists to prevent themselves from drowning. Finally, local leaders produced their own Potemkin villages to demonstrate success where there was none. Newspapers published pictures of children able to sit on wheat because it was so dense. Investigations later determined that the pictures were faked by putting a bench beneath the children.[131]

Mao also dramatically increased the pace and scope of infrastructure development for agriculture, including a massive effort to build reservoirs, dams, and irrigation projects: "Every county in China was ordered to construct a water reservoir by building a dam and water channels. . . . Most of the county dams had collapsed within two or three years and the dam on the Yellow River [Sanmenxia] quickly filled up with silt, rendering it next to useless."[132] Such waterworks projects necessitated huge forced relocations. In Zhejiang Province, 300,000 people were relocated from one county alone to make room for the Xinanjiang reservoir.[133]

Yet another environmentally devastating innovation of the Great Leap Forward was the campaign to produce iron and steel in backyard steel furnaces in an effort to surpass the industrial achievements of the West. Mao threw down the gauntlet: "In another two years, by 1962, it is possible [for us to produce] eighty to a hundred million tons [of steel], approaching the level of the United States. . . . [At the end of] the second-five year plan, we will approach or even surpass America."[134] As Becker notes, "The entire country, from peasants in remote villages on the Tibetan plateau to top Party officials in Zhongnanhai in Beijing, set up smelters in 1958 and 1959 to create 'steel' in backyard furnaces. Everyone had to meet a quota by handing over their metal possessions . . . bicycles, railings, iron bedsteads, doorknobs, their pots and pans and cooling grates. And to fire the furnaces, huge numbers of trees were cut down."[135]

The result was a massive amount of useless metal and skyrocketing pollution levels. Beijing was transformed from a city that "did not produce even pencils" to one that boasted "700 factories and 2000 blast furnaces belching soot in the air."[136] All told, the Great Leap Forward yielded "600,000 shabby iron and steel making furnaces, 59,000 coal mines, 4000 power stations, 9000 cement factories and 80,000 farm repair shops. The number of factories increased from 170,000 in 1957 to 310,000 in 1959."[137]

Statistics bear out this period's devastating impact on the environment. One estimate held that to implement Mao's policy of backyard steel furnaces, for example, as much as 10 percent of China's forests were cut down.[138] Qu notes,

> The environmental situation quickly deteriorated. A lot of places were polluted by either smog, sewage waters or rubbish. Under a zealous drive, mineral resources were also exploited, resulting in startling losses and destruction to both topography and landscape. Biological resources, forests in particular, were seriously damaged, causing several losses to the ecosystem. There was extensive destruction of the natural environment of our country.[139]

Perhaps nothing could exceed the horror, however, of the massive famine wrought by the excesses of the Great Leap Forward. The death toll reached an estimated 35–50 million people during 1959–61;[140] within one year alone (1959–60), China's population shrank by 10 million.[141] In 1960, infant mortality reached 330 per 1,000.[142]

Cultural Revolution (1966–76) and Its Aftermath

In the wake of the devastation wrought by the Great Leap Forward, Chinese leaders again began the slow process of rebuilding the social and economic infrastructure of the country. In 1966, only five years after the policies of the Great Leap Forward had been modified or repudiated, Mao initiated a second "revolution," with equally devastating consequences for the country and the environment.

The Cultural Revolution wrought even greater devastation on the economy and the social fabric of the country than the Great Leap Forward, with long-term negative implications for the environment. Qu comments, "The few environmental regulations in industry, agriculture and urban constructions were repudiated and negated as bourgeois and revisionist restrictions. Cases of environmental and ecological damages rapidly increased to a terrifying degree."[143]

Several policies associated with the Cultural Revolution contributed to the rapid destruction of the natural environment. Industrial production stressed quantity and high output with little concern for using appropriate technologies. Thus there was significant waste in raw materials and energy. Labor and environmental regulations were discarded, contributing to substantial increases in air and water pollution along with significant loss of biodiversity.[144]

The Cultural Revolution brought a renewed emphasis on grain production to the exclusion of forestry, animal husbandry, and fisheries. Forests and pastures were destroyed, lakes were filled, and man-made plains were cultivated to grow grains. One such campaign to expand the amount of cultivated land in northern Manchuria and Xinjiang was bound to fail since little could be done to "lengthen the north Manchurian growing season or ameliorate its soils, except at prohibitive cost."[145]

Vaclav Smil details the degradation that occurred to the land during the Cultural Revolution:

> Illegal felling of forest trees, always a problem in wood-short China, became truly rampant, and the worsening rural energy supply intensified the damage. . . . In most places the inevitable vicious circle set in soon after slopes were deforested to make way for grainfields: after a few years, as the accumulated organic matter was sharply reduced and the thin soil rapidly eroded, yields on the newly reclaimed land plummeted and more land was deforested just to maintain the harvests. The abandoned, barren land then

succumbed to erosion, often with the irreparable result of all soil being removed to the bedrock.[146]

Mao's "third front" policy, which called for moving production away from the coastal regions to protect against an attack from a foreign power, led to the siting of factories "near the mountains, scattered and into the caves." As a result, officials constructed factories spewing toxic discharge in the mountains, polluting the atmosphere and water.[147]

Yet, as with the Great Leap Forward, the most destructive consequences of the Cultural Revolution were exerted on the whole of Chinese society. China's educated elite—and by extension the country's research and intellectual advancement in all fields, including environmental protection—were devastated. Tens of thousands of intellectual leaders and their families were sent to labor in the fields or factories. Others were imprisoned or killed. In the fifteen or so years spanning both the Great Leap Forward and the Cultural Revolution, Mao had easily equaled the worst excesses of Imperial China, tearing apart the social fabric of the country, devastating the economy, and ravaging the environment.

The Legacy

China's history suggests a long, deeply entrenched tradition of exploiting the environment for man's needs, with relatively little sense of the limits of nature's or man's capacity to replenish the earth's resources. Attitudes, institutions, and policies were rooted in and supported by traditional concepts and philosophies such as Confucianism that promoted man's need to overcome nature in order to utilize it for his own benefit, while the relatively eco-friendly philosophies of Taoism and Buddhism made limited inroads in the consciousness of the Chinese people and leaders.

In practice, the relentless drive to amass power, develop the economy, and meet the basic needs of a burgeoning population led to the plundering of forests and mineral resources, poorly conceived river diversion and water management projects, and intensive farming that degraded the land. The continual cycles of social transformations (including war, population growth, and economic development) and eco-environmental change resulted in large-scale deforestation, desertification, soil erosion, and flooding that only increased in scope and scale over time.

China's institutional capacity to protect the environment was strictly personal, with responsibility for maintaining harmony between man and nature resting primarily in the hands of the emperor or leader of the country and secondarily in the offices of regional administrators. Simply put, the emperor's "mandate from heaven" coupled with a Confucian-supported reliance on officials to behave morally and responsibly in executing their duties served as the mainstay of environmental protection for much of China's history. Thus, how effectively land and water resources were protected and conserved depended overwhelmingly on the proclivities of individual officials. While some enlightened statesmen and provincial officials, such as Guan Zhong and Yu Sen, advanced environmentally sound policies, most sought to develop the economy or military as rapidly as possible, giving little thought to the limits of the environment's resources.

At the same time, the general populace played virtually no role as advocate for environmental protection. Exceptions occurred in instances where their immediate environment and resources were threatened. Polluted water or damaged crops, for example, might incite serious local conflict. But for the average person in China, environmental protection was the purview of central and regional officials; only when immediate interests were threatened did a citizen take action.

The weakness of this personalistic system of environmental protection was compounded by a poor tradition of codification and enforcement of environmental laws. The blend of Confucian values and Legalist ideals reinforced the power of the local magistrate in determining on a case-by-case basis how best to balance competing interests rather than relying primarily on a set of codified laws. This greatly enhanced the ability of the emperor or local officials to act in an arbitrary manner with overwhelmingly negative ramifications for the environment. By extension, institutions devoted to the enforcement of codified law also were weak or nonexistent. Finally, the lack of a strong legal infrastructure enabled corruption to flourish within Chinese society, further confounding environmental protection efforts and contributing to social dislocation and turmoil.

Moreover, China's history suggests a remarkable consistency in the methods that the country's leaders adopted to manage their natural resources. From emperors to Mao, China's leaders favored campaign-style mass mobilizations for resolving environmental problems such as flood control or deforestation. But these campaigns often failed. In emphasizing the pace and scope of the effort, little consideration was given to the actual environmental and scientific factors necessary to achieve success, and without the freedom to question scientific beliefs and practices and to propose alternatives, the Chinese expert community was stifled in its ability to provide informed and useful analysis to the political elites.

The legacy of China's traditional environmental attitudes, institutions, and policy approaches therefore provided little foundation for building a sound environmental protection apparatus. The first decades of China's post-Mao leadership created new and frightening threats to the natural environment.

THE ECONOMIC EXPLOSION AND ITS ENVIRONMENTAL COST

The death of Mao Zedong in 1976 and the end of the Cultural Revolution opened the door to a fundamental shift in China's political position both domestically and internationally. Beginning in 1978, the Chinese leadership embarked on a reform program that continues today and has taken the country into uncharted territory. Increasingly, the market rather than the state drives the Chinese economy. There is far greater freedom to travel, speak openly, and engage in private social activities, and China has become integrated into the international community through trade and participation in an array of international organizations and agreements.

This reform process has also created a new set of environmental problems. Deng Xiaoping's early 1980s call to arms, "To Get Rich Is Glorious," set the stage, if inadvertently, for yet another state-sponsored campaign to exploit the natural environment for the purpose of economic development. Today, more than two decades of economic development, scarcely restrained by nascent environmental protection institutions, have reinforced China's tradition of promoting rapid economic growth at the expense of the environment.

At the same time, the Chinese leadership has taken a number of important steps toward balancing the needs of a burgeoning population with the country's natural resources. In an effort to reduce the population growth rate, it has established an overarching family planning bureaucracy and adopted a range of restrictive population regulations. While successful by many measures (as discussed below), China's population policy has been undermined by the very economic reforms it has sought to enhance.

The state must also contend with secondary social and economic challenges engendered by the interplay of economic reform and the environment. What are these challenges? At the top of Beijing's agenda are the wide-scale migration from rural to urban areas and the social tensions that have ensued; growing instability in rural and urban areas; increasing public health problems; and the significant costs to China's economy of environmental pollution and degradation in terms of lost work days, factory shutdowns, and remediation projects, such as cleaning up China's already highly polluted lakes and rivers. As environmental challenges continue to multiply, moreover, the policy measures necessary to respond effectively increase in scope and complexity, further increasing the economic and political burden on China's local and central leaders.

Economic Miracle

The reform of China's economy is one of the great success stories of the last quarter century. From the late 1970s on, pressures from below and initiatives from above have transformed an economy crippled by decades of state control and poorly conceived state-sponsored development campaigns into a global economic powerhouse.

One key factor was the central government's decision in the early 1980s to devolve authority for economic development to the provinces, while conferring provincial status on a number of cities. Provincial leaders gained substantial fiscal authority, the ability to approve capital construction projects and foreign joint ventures, and greater leeway to appoint officials.[1] The results were striking: "Local government and CCP [Chinese Communist Party] leaders responded to the new incentives with a burst of entrepreneurial energy. They founded new local industries and pitched the merits of their provinces to foreign investors. Stimulated by local initiatives, local and national growth rates skyrocketed."[2] Gross domestic product (GDP) per capita increased more than tenfold from approximately $84 (692 yuan) in 1984 to $1100 (9115 yuan) in 2003.[3]

In the countryside, pockets of such economic dynamism were evident even before Deng's reforms. With the complicity of local officials, farmers in some provinces had been experimenting with private plots and sideline enterprises. In 1983, Deng and his supporters explicitly acknowledged the farmers' achievements, and under the rubric "household responsibility system," sanctioned countrywide implementation of the reforms. Grain production in many cases doubled.

The farmers' economic success and growing financial autonomy spurred other far-reaching reforms of the Chinese rural and urban economies.[4] The farmers established a de facto labor and grain market, thereby undermining the system of grain rationing (*hukou*) that the state had established to keep farmers tied to the countryside. By the early 1990s, it was possible to buy grain openly in farmers' markets or through the black market. This led huge numbers of rural migrants to seek work in China's dynamic coastal cities, providing much needed low-end services but also stretching the resources of many of these cities.

The central leaders also began to plan, as early as 1987, to diminish the role of the state-owned enterprises (SOEs) in the Chinese economy. Beginning in the early 1960s, in many cities, these massive SOEs—sometimes employing upward of one hundred thousand people—were the bread and butter of the local economy. They dominated the key infrastructure sectors of the economy: power generation, ferrous metals, railroads, chemicals, machinery, and even textiles. Typically, they employed tens of thousands of people and provided for all the social welfare needs—education, health, and retirement—of their workers. Like their Soviet counterparts, however, they were by and large notoriously inefficient money sinks for state capital. A decade later, in 1997, Premier Zhu Rongji began to push aggressively to dismantle the system, an economically and politically challenging effort that has produced substantial unrest in many urban areas.

Since then, in place of SOEs as the foundation of the urban economy, the Chinese leadership has banked on collective, private, and joint venture enterprise. In the countryside, township and village enterprises (TVEs) have been widely touted as a key engine of economic growth capable of absorbing much of the excess labor in the countryside and encouraging farmers to remain on the land rather than migrate to the cities. By 2000, there were 20 million TVEs with an estimated 128 million employees, responsible for 30.4 percent of China's GDP and 34 percent of rural income.[5]

Foreign investment and joint ventures with Chinese enterprises have played an equally significant role in the new economy. In the early 1980s, the Chinese leadership designated some cities and provinces, primarily along the coast, as Special Economic Zones. As a result, total foreign direct investment skyrocketed from $430 million in 1982 to $53.5 billion in 2003.[6] The standard of living in the coastal areas similarly benefited; per capita GDP in Shanghai, for example, jumped from approximately $300 in 1978[7] to $4,950 in 2002.[8] International governmental organizations (IGOs) like the World Bank and the Asian Development Bank began to play a substantial role in developing infrastructure such as highways, railroads, ports, and energy projects as early as the mid-1980s.

As China's leaders survey the results of twenty-five years of reform, they have cause for great pride. Economic reform has brought revolutionary change to the lives of many Chinese citizens. Increasing standards of living, better access to goods and services, and greater freedom of movement and job choice all have reshaped the opportunities and life prospects of hundreds of millions of Chinese.

Environmental Havoc

The same dynamic that produced such success in the economic sphere, however, has also wreaked havoc on China's natural environment. The burgeoning economy has dramatically increased the demand for resources such as water, land, and energy. Forest resources especially have been depleted, triggering a range of devastating secondary impacts such as desertification, flooding, and species loss.

Environmental protection efforts have lagged. In the mid-1970s, the central government began to develop a small-scale environmental protection effort that gradually increased in scope and power. Local environmental protection bureaus, however, are wholly dependent on local officials for their economic welfare, and the central State Environmental Protection Administration (SEPA) has long struggled to make itself heard within the overwhelmingly development-oriented top leadership.

Without a strong, independent environmental protection apparatus, the devolution of authority to provincial and local officials has given them free rein to concentrate their energies on economic growth, pushing aside environmental considerations with few consequences from the center. Thus, in many regions, land, water, and forest resources have been squandered, without considering the necessity of conservation or replenishment of these natural resources.

The small-scale TVEs that have fueled much of China's growth have proved difficult to monitor and regulate. As they have increased in importance to the Chinese economy, they have rapidly proved themselves an equal, if not greater, threat to the environment than the SOEs. By 2000, these TVEs were estimated to be responsible for 50 percent of all pollutants nationally.[9] Even the most committed environmentalists, like Xie Zhenhua, the head of China's SEPA, tacitly recognize the primacy of economic imperatives in Chinese policy. When calling for heightening control on pollution by township enterprises, Xie simultaneously promised that environmental regulation would not hamper rural industrial development.[10] And in December 2000, Xie granted more than five hundred SOEs that had been facing closure a two-year reprieve to meet pollution standards.[11]

Integration with the global economy, while providing some environmental benefits, has also contributed to China's new status as a destination of choice for the world's most environmentally damaging industries—petrochemical plants, semiconductor factories, and strip mining among others—and provided an insatiable global market for China's resource-intensive goods such as paper and furniture.

The results of decades of this interplay between the economy and the environment have been devastating. In many ways, the reform process is leaving as large a footprint on the natural environment as did centuries of imperial, republican, and early Communist rule.

Felling China's Forests

Within the first several years of economic reform (1978–1986), as local officials raced to take advantage of the new economic incentives and the relative lack of regulation, logging increased by 25 percent.[12] By the mid-1990s, local officials reported that of the country's 140 forest bureaus, 25 had exhausted their reserves and 61 had indicated that trees were being felled at unsustainable rates.[13] Highly skilled and productive lumberjacks became local heroes, felling trees as swiftly as possible to meet the growing domestic and international demand for China's timber products. Throughout the 1980s and 1990s, China's timber production skyrocketed as demand for chopsticks, furniture, and paper[14] drove an increasingly profitable legal and illegal logging business. China's integration into the international economy also led to an influx of Japanese and Taiwanese multinationals into China's logging industry. China, in turn, was able to tap into a lucrative international market for its wood products. China is now the world's second largest consumer of timber.[15]

In a country where centuries of demands for fuelwood, cropland, and war had already reduced per capita forest reserves to one of the lowest levels in the world, the forests were ill-equipped to survive this new onslaught.[16] Today, China reports forest coverage of 16.55 percent,[17] well below the U.S. coverage of 24.7 percent and the world average of 27 percent.

Reports from individual provinces highlight the dramatic decline in forest coverage over the past few decades. In Sichuan, for example, one of the most heavily forested provinces in the country and home to the country's famed pandas, the ratio of trees felled to trees planted through the 1980s and mid-1990s was as high as ten to one. The local inhabitants were said to be "replacing the forest of trees with a forest of arms and axes,"[18] and forest coverage in the province dropped from 28 percent in the 1970s to 14 percent in the 1980s. By the late 1990s, the province was left with just 8 percent of its original trees.[19] Sichuan wasn't alone. In the upper reaches of the Yangtze, forest coverage dropped from 30–40 percent in the 1950s to only 10 percent in 1998.[20] And, in several provinces, autonomous regions, and municipalities, such as Qinghai, Xinjiang, Shanghai, Ningxia, Tianjin, Jiangsu, and Gansu, forest cover dropped below 4 percent.[21] Even the worst examples of deforestation in the United States, such as Vermont's transformation from 70 percent forest to 30 percent over the past century, are mild in comparison to China's experience.[22]

While China's forest resources have suffered substantial loss, so too have China's grasslands. Grasslands now account for about 40 percent of China's territory, primarily in the western areas of Tibet, Xinjiang, Qinghai, and Inner Mongolia. Overall, degradation has reduced China's grasslands by 30–50 percent since 1950; of the 400 million or so hectares of natural grasslands remaining, more than 90 percent are degraded and more than 50 percent suffer moderate to severe degradation, contributing to decreased biodiversity and diminished capacity to serve as watershed protection. Annual reports from SEPA indicate no improvement in the situation since 1996.[23]

Beginning in the 1950s and continuing until today, in an effort to boost China's domestic grain production, millions of hectares of grasslands have been converted to irrigated crop production, leading to severe degradation of the land. The reform period has further introduced new challenges to efforts to protect China's grasslands: privatization of herds and grazing land, intensive grazing management strategies, and new farming techniques for growing forage and fodder have all undermined grassland protection.[24]

The loss of China's forests and degradation of its grasslands have both local and global consequences.[25] Domestically, China suffers from wood shortages, altered ecosystems, soil erosion, riverbed deposits, flooding, and changing local climates. Globally, deforestation contributes to climate change through the release of carbon dioxide when trees are felled, and from the loss of a carbon sink. Deforestation, along with the degradation of the grasslands and over-cultivation of cropland, has also contributed to the growing desertification and increasing number of devastating sandstorms that are transforming China's north.

More than one-quarter of China is now desert. In the northwest, the pace of desertification more than doubled from 1,560 square kilometers (sq km; approximately 600 square miles) annually in the 1970s[26] to 3,436 sq km (approximately 1,300 square miles) annually in the latter half of the 1990s,[27] producing a continuous stream of migrating farmers and herders. As the Chinese say, "The desert marches on while human beings retreat."[28] In May 2000, then premier Zhu Rongji worried that the rapidly advancing desert would necessitate moving the capital from Beijing, although assessments by China's scientific community suggest that such a dire outcome is unlikely to result.

By the late 1990s, an average of thirty-five sandstorms was also wreaking havoc in northern China every year, compared with fewer than twenty two decades ago.[29] Beijing has been repeatedly blanketed in dust from desert sandstorms, "obscuring the sun, reducing visibility, slowing traffic, and closing airports,"[30] and in which "thousands of kilometers of highways and railroads are blocked by sedimentation."[31] During spring 2000, Beijing was hit with eleven such sandstorms, and in April 2001, the dust was so severe that it traveled not only through much of East Asia but also reached Canada and the United States, as far east as New England and as far south as Florida.[32]

In addition, logging, loss of grasslands, wetlands reclamation, and pollution directly threaten China's vast biodiversity: Of the 640 species listed by the Convention on International Trade in Endangered Species, 25 percent are found in China, and 15–20 percent of the plant and animal species in China are endangered.[33] China's efforts to develop nature reserves to stanch this loss are impressive. As of 1997, China boasted almost 1,000 nature reserves throughout the country. Still, only two-thirds of the reserves actually had staff and budgets, and of those, many were forced to resort to commercial ventures within the reserves to raise money.[34]

Occurring over decades, if not centuries, the magnitude of China's deforestation and desertification has been difficult for the leadership to recognize and address. In 1998, however, nature sent the Chinese leadership a wake-up call. The great Yangtze River, stretching from the Tibetan plateau to the East China Sea, flooded, killing more than 3,000 people, inundating 52 million acres of land, and causing $20 billion in economic damages. Rampant logging, along with destruction of wetlands, which eliminated the natural capacity of the land to absorb floodwaters, was deemed the primary culprit. Then premier Zhu Rongji immediately banned logging for huge swaths of western Sichuan Province. The ban has been extended to seventeen provinces, autonomous regions, and municipalities. At the same time, the government has announced a campaign with an estimated cost of $725 million to prevent and control desertification by adding new grassland and forest and increasing the vegetated area throughout the northwest. The government is also directing some resettled farmers and herders to change their sources of livelihood from agriculture to tree planting. Yet, the success of the campaigns is in doubt, as we'll explore in chapter 4.

The Search for Water

In 1998, in Taiyuan, Shaanxi Province, farmers unscrewed manhole covers from sewage pipes to irrigate their crops. Residents survived for days without tap water; when water was provided, it was sometimes available for only an hour a day. Moreover, the major coal mine in the region, Datong, was losing $100 million in revenues annually because it did not have enough water to wash its coal. Confronted with this situation, officials in the city faced three unattractive and expensive options: (1) relocate the two million or so residents of the city; (2) close all heavy industry; or (3) undertake a costly river diversion project. They elected to pursue the third alternative, bringing the province into ongoing conflict with its similarly water-poor neighbor, Qinghai.

For many regions in China, diminishing water supplies pose today's greatest social, economic, and political challenge. At 2,500 cubic meters (m³) per capita, China's national supply of freshwater is well above 2,000 m³, the World Bank's definition of a water-scarce country.[35] However, this figure does not account for the substantial regional disparities in water access. Water distribution is highly uneven, with availability greatest in the south and much less in the north: average rainfall is approximately nine times more in the southeast (1,800 millimeters [mm]) than in the northwest (200 mm). Over 45 percent of China receives less than 400 mm precipitation a year. The distribution of groundwater resources is similarly skewed: average groundwater deposits in the south are over four times greater than in the north.

It is no surprise, therefore, that officials in Shanghai, Guangzhou, and Taiyuan have all cited water scarcity as their number one environmental concern, and the Ministry of Water Resources predicts a "serious water crisis" in 2030, when the population reaches 1.6 billion and China's per capita water resources are estimated to decline to the World Bank's scarcity level.[36] Already, about sixty million people find it difficult to get enough water for their daily needs,[37] and more than ten times that many drink water contaminated with animal and human waste.[38]

Rapid economic growth has significantly increased demand for water in the agricultural, residential, and industrial sectors of China. Most of China's water is directed toward agriculture; in northern China, for example, 85 percent of arable land is irrigated. (In contrast, only about 10 percent of U.S. arable land is irrigated.[39] Although in California, where demand for water often exceeds supply and is beginning to constrain the development of new housing projects, the agricultural sector uses 80 percent of the state's water.)[40] Demand for water is growing at an annual rate of 10.1 percent by cities and 5.4 percent by industry.[41] In some of the most dynamic regions of the country, demand is even greater. From 1988 to 1998, daily per capita consumption of tap water in Shanghai increased by more than 25 percent;[42] in an even shorter period (1988–1994), demand for water in Guangzhou rose by 35 percent.[43] Such trends show no sign of abating. Scientists predict that with its growing industrial base, Guangzhou's water demand will double between 2000 and 2010.[44]

Years of drought have also depleted China's water resources. Baiyangding Lake, previously considered the "pearl of North

China," is now dry, and hundreds of thousands of fishermen weave mats for their livelihood instead.[45] The Yellow River, once referred to as "China's sorrow" because its high waters caused so much destruction, has been running dry in places since 1985. In 1999, it ran dry for forty-two days,[46] disrupting shipping and commerce. By 2001, Chinese scientists reported that more than two thousand lakes in Qinghai that nurture the Yellow River were disappearing due to climate change and overutilization. Not only has this affected the water level in the river itself but it has also diminished the region's agricultural, industrial, and energy (from hydropower) output.[47] The Huai River and Hai River, also in northern China, are also considered especially vulnerable to serious water shortages.[48] Along some areas of the Hai, the groundwater level has fallen 50 meters.[49] Even in the traditionally water-rich Yangtze Valley, fifty-nine cities suffered from inadequate water supply in 2002, and officials fear that serious problems for water transport will soon arise.[50]

As demand for water has skyrocketed, so, too, have levels of water pollution. According to SEPA's annual report for 2003, over 70 percent of the water in five of the seven major river systems—the Huai, Songhua, Hai, Yellow, and Liao—was grade IV or worse (not suitable for human contact). Almost 30 percent of the water in the monitored sections of the Yangtze was also grade IV or worse. Only the Pearl river system provided more reassuring statistics: more than 80 percent of the water tested reached grade III or better (suitable for human contact). Moreover, despite a large-scale cleanup campaign for three of these rivers and three of China's largest lakes (*san he san hu*), water quality since the mid-1990s has either remained the same or deteriorated further.[51] In the three lakes—the Tai, the Hu, and the Dianchi—the 2003 statistics provided by SEPA indicate that more than 70 percent of the monitoring stations for the Tai lake report water quality of grade V (suitable only for irrigation) or worse, while all the monitoring stations for the Hu and Dianchi lakes report water quality of grade V or worse.

Perhaps most striking, only six of China's twenty-seven largest cities supply drinking water that meets state standards.[52] The challenge of providing clean drinking water is exacerbated by the steadily increasing amount of industrial and municipal wastewater discharge. Reporting for 2002 indicated an increase of 1.5 percent in the amount of industrial and municipal wastewater; and during 2002–2003, combined industrial and municipal wastewater amounts increased by 4.7 percent. In addition, treatment of wastewater does not always lead to the attainment of government standards. Still, in 2002, SEPA reported that 89.4 percent of industrial wastewater from specially targeted enterprises reached national standards, while 73.9 percent of industrial wastewater from normal enterprises reached national standards. (Treatment rates for municipal wastewater were not provided.)[53]

Outside China's major cities, the rise of TVEs has dramatically increased local water pollution. Factories and municipalities dump their untreated waste directly into streams, rivers, and coastal waters.[54] The proliferation of tanneries, chemical and fertilizer factories, makers of brick, tile, pottery and porcelain, small coal-fired power plants, and pulp and paper factories have all contributed to a dramatic increase in pollution outside China's major cities. By one estimate, TVEs discharge more than half of all industrial wastewater in China.[55] As one analyst has noted,

> Enormous numbers of small to medium sized rural and township enterprises have proliferated outside China's large cities in recent years. In Jiangsu province, which surrounds Shanghai, roughly one such enterprise can be found per square kilometer. But along with jobs these plants have brought a variety of water pollutants into the Chinese countryside. Unknown volumes of untreated waste from these plants get dumped into streams and networks of canals, contaminating the water used for drinking, irrigating fields, and watering animals.[56]

Regulating these firms is difficult. Although the National Environmental Protection Agency announced a crackdown on small paper mills in 1994, for example, Chinese scientists stated at the time that they had no means of controlling the pollution from these small-scale enterprises.[57] In some cases too, local residents have accepted the pollution generated by the TVEs because they depend on the factories for jobs.[58] Complicating the problem is that roughly 200 million people live in towns that possess no sanitation system other than "pipes that lead wastewater to the nearest ditch."[59]

Less visible but equally insidious has been the pollution generated from excessive use of chemical fertilizers. Indeed, China's use of chemical fertilizers has more than quadrupled during the reform period, from 8,840,000 tons in 1978 to 42,538,000 tons in 2001.[60] According to the World Bank, the poor quality of fertilizers and their inefficient application is contributing to significant nutrient runoff, which in turn is contributing to eutrophication in many of China's most important lakes, in which the growth of dense algae depletes the shallow water of oxygen.[61]

Water scarcity and pollution harm the Chinese economy by desiccating or polluting cropland, forcing investment of valuable domestic resources into large river diversion projects and cleanup efforts, and contributing to growing social tensions in both rural and urban areas. And there is little relief in sight. Chinese scientists have predicted that by 2020, water shortage may exceed 50 billion m^3, more than 10 percent of the country's total current annual consumption.

Fueling China's Economy

The most visible sign of environmental pollution in China is the thick haze that periodically settles over cities around the country. In December 1999, then premier Zhu Rongji commented to Beijing city officials, "If I work in your Beijing, I would shorten my life at least five years. . . . Every year, after we start heating in the winter, Beijing's atmospheric environment deteriorates. Foul air from burning coal and car exhausts cannot disperse easily and forms a thick 'pan cover' over Beijing. Sometimes when passengers coming to Beijing ask air hostesses, 'Where is Beijing?' they simply answer: 'Under the cover of the pan.' "[62] Or, as one reporter has noted, "On the worst days, smog is still so thick in the capital that it can render a 50 story building invisible from 100 yards."[63]

Beijing is one of sixteen Chinese cities among the twenty world cities with the most polluted air.[64] In 2002, SEPA tested the air quality in more than three hundred Chinese cities and found that almost two-thirds failed to achieve standards set by the World Health Organization (WHO) for acceptable levels of total suspended particulates (TSPs), which are the primary culprit in respiratory and pulmonary diseases. In addition to these ultrafine particulates, sulfur dioxide (SO_2) emissions, which cause acid rain, are now the highest in the world, affecting over one-fourth of China's territory, including 30 percent of China's agricultural land, mostly in southern and central China.[65] Acid rain poisons the country's fisheries, ruins cropland, and erodes buildings. According to one estimate, by 2002, acid rain caused almost $13.3 billion in damage to human health, farms, and forests in China annually.[66] Japan and South Korea also blame China for much of their problems with acid rain, a situation that has contributed to ongoing tensions in the region.

The rampant air pollution stems, in significant measure, from China's continuing reliance on coal to supply more than two-thirds of its energy needs. (By contrast, in Japan, the United States, and India, coal accounts for almost 17 percent, 23 percent, and 51 percent, respectively.)[67] Oil represents an estimated 23.6 percent of China's total energy consumption. Cleaner energy such as natural gas and hydropower, in contrast, accounts for only 2.5 percent and 6.9 percent, respectively, of China's total energy consumption.[68] One positive environmental trend is the steady expansion of coal gas and natural gas for district heating in urban areas: since 1985, their use has increased more than five times.[69]

Burning coal is responsible for 70 percent of the smoke and dust in the air and 90 percent of the sulfur dioxide.[70] The economic reforms have only worsened the problem. Over the course of the reform period, China's coal use has doubled from just over 600 million metric tons to more than 1.2 billion metric tons, making it the world's largest consumer of coal.[71]

The sources of this coal burning are diffuse and often difficult to control. Most troublesome are the inefficient industrial boilers used in outdated factories and power plants, as well as small household stoves used by the majority of Chinese.[72] Well over half of the Chinese population uses coal in their homes, and about one-fifth of rural homes rely on coal for domestic fuel.[73]

The central government has tried to consolidate energy suppliers into large-scale power plants, where environmental technologies are easier to employ and enforce. But local governments have encouraged the proliferation of small (under 50 megawatts [MW]), inefficient, and highly polluting coal-fired power plants to meet growing local energy needs. The smallest of these plants (12 MW) release three to eight times more particulates, consume 60 percent more coal,[74] and are 35–60 percent more costly to operate than plants of 200 MW or more.[75] At the same time, small coal mines have sprouted throughout coal-rich areas, prompting additional health, safety, and environmental concerns. Campaigns to close down these small-scale, local plants and mines have met with mixed success.[76]

China's integration into the world economy has been a dual-edged sword with regard to the country's air quality, especially in southern China. While many multinationals have significantly elevated the level of environmental technology employed in Chinese enterprises (as explored in chapter 6), others, with the complicity of local officials, have taken advantage of China's weaker laws and enforcement capacity to relocate their most polluting enterprises to the mainland. For example, in the late 1990s, SEPA accused Taiwanese and South Korean multinationals of establishing their factories in China in order to avoid stricter domestic environmental regulations.[77]

Hong Kong businesses, especially, have taken advantage of weaker environmental laws and enforcement on the mainland to site their highly polluting industries in nearby Guangdong province. In the early 1990s, many Hong Kong businesses relocated there to take advantage of lower wages and to avoid a ban on sulfur-heavy fuel for industrial use.[78] Yet Hong Kong itself has begun to pay the price for shipping these factories across the Pearl River. Guangdong province produces about 690,000 tons of sulfur dioxide annually (compared to Hong Kong's 80,000 tons), and in October through April the wind blows sulfur and nitrogen dioxides from these factories into Hong Kong, producing noxious and poisonous "cloudbanks" over the island. In the words of one mainland engineer, "Hong Kong companies use us to make money, but in the end what they do goes back to haunt them."[79] Other countries have begun to dump their toxic waste from the high-tech sector in China. Hong Kong and Taiwanese brokers, for example, buy hazardous electronic scrap from the United States and other countries to recycle in China. Whatever is of value is sold; the rest is typically burned and dumped, fouling the air and polluting China's lakes and rivers. In one case, in Guiyu, Guangdong Province, at a site where circuit boards had been processed and burned, levels of lead in the water were 2,400 times higher than WHO drinking guidelines,[80] and heaps of black ash dotted the area.

Hong Kong officials are greatly concerned about the impact of the mainland's environmental practices on the island's water and air quality and have established a number of joint working groups and collaborative efforts with various arms of the mainland government, including the State Oceanic Administration and the Guangdong provincial government. Yet success is elusive. As one Hong Kong official stated, "We might have more stringent standards than Shenzhen, but we cannot enforce them across the boundary or ask them to do more to meet our standards."[81]

The greatest future challenge to China's air quality—both on the mainland and in Hong Kong—will likely arise from a new source: the country's rapidly growing transportation sector. Currently, there are only about 13 vehicles per 1,000 persons in China,[82] compared to the world average of 114 vehicles per 1,000 persons. But the country's fleet of cars, trucks, and buses is growing by 13 percent per year, and in major cities that figure rises to 20 percent.[83] The number of officially registered vehicles has risen from 1,358,400 in 1978 to 3,496,100 in 1985 to 25,727,200 in 2001.[84] Already, major cities such as Beijing, Guangzhou, and Shanghai suffer from severe traffic gridlock. A trip from one part of downtown Beijing to another that took ten minutes fifteen years ago now takes a half hour or more. By 2050, if China were to match the United States, with one of every two Chinese owning a car, the country would boast 600 to 800 million cars, a number equal to today's world total.[85]

Domestically designed and manufactured cars pose China's most significant threat to air quality. These cars emit 10–20 times more carbon monoxide, nitrogen oxide, and hydrocarbons than U.S. and Japanese models. Poor maintenance worsens these levels several-fold.[86] In 2000, Beijing boasted 1.5 million vehicles,[87] roughly one-tenth the total in Tokyo or Los Angeles—yet the pollution generated by the vehicles in Beijing equaled that of the other two cities.[88] While China's entry into the World Trade Organization and stringent new fuel efficiency standards announced in 2003 may increase the role of foreign car companies in shaping China's transportation future, thereby improving air quality prospects, analysts expect that a vast portion of the Chinese population will continue to be serviced by lower-cost and more polluting Chinese vehicles.

Economic Reforms and Population

The expected growth in China's transportation sector highlights the importance of China's population size as a continued environmental challenge for the Chinese government. In the waning years of Mao's tenure, China's leaders began to move away from the centuries-old notion that the country's future greatness was premised on a large population base. The first sign of this policy shift appeared in the early 1970s, when Premier Zhou Enlai proposed that population planning be incorporated into the state plans for the development of the national economy. During 1970–73, a government slogan said, "One child will do, two are good enough and three are one too many." From 1974 to 1977, this was revised to "late marriage, fewer births and wider spacing," encouraging one birth and two at most. Encouraged by Beijing, the provinces implemented population control programs, cutting the birth rate in half.[89]

The ascension of Deng Xiaoping and a reform-oriented leadership in 1978 broadened the scope and accelerated the pace of this policy shift. In 1979, convinced of the necessity of further slowing the population growth rate to achieve economic progress, the National People's Congress approved the one-child policy.[90] This cemented the leadership's commitment to maintaining a low population growth rate. It also revived some of the more draconian policy prescriptions that previous population planning enthusiasts, such as Qing scholar Wang Shiduo, had advocated. Compulsory abortion, for example, became prevalent in areas with aggressive family planning officials.

By all accounts, China's leaders have succeeded in slowing the rate of population growth from the prereform period. However, family planning efforts achieved their most dramatic impact early on, with less consistent results as the reforms progressed. Thus, despite official claims of success, during a European tour in October 1999, then president Jiang Zemin acknowledged that the biggest problem he confronted was China's population size.[91]

Paradoxically, the very economic success to which the one-child policy has contributed has also served to undermine the one-child policy in much of rural China. The economic independence of China's farmers, born of the reforms, has enabled many of them to ignore this policy.[92] Despite frequent adjustments to the incentive structure and the implementation of coercive policies (such as financial penalties or even forced abortion), farmers have managed to resist the strict implementation of the one-child policy. Farmers who have benefited from the reforms in the rural economy and the new opportunities to migrate to large cities for higher paying work, for example, are often willing to pay the fines imposed for having more than one child or to bribe local officials to look the other way. Jiang himself pointed to local corruption as a key source of family planning failure in the country. Midwifery in which a birth might not be officially documented, underground railways that move pregnant women to other towns or cities, and migration are other options for farmers and their families.[93]

In addition, in provinces heavily populated by ethnic minorities, such as Tibet, Yunnan, and Xinjiang, family planning policies were left largely to the discretion of local leaders to work out "in line with their realities."[94] As a result, these populations expanded far more rapidly than those in the rest of the country.

In late March 2001, in the wake of the Fifth National Census, the government claimed victory in its family planning efforts. The population had reportedly peaked at 1.295 billion, thereby apparently just meeting the country's target of keeping the population below 1.3 billion by the end of 2000,[95] although it exceeded the target set by Premier Li Peng in 1988 of 1.2 billion people by the end of the century.[96]

Some Chinese scholars, however, have questioned the accuracy of the 2001 census,[97] and many are concerned with the broader social, economic, and environmental implications of China's demographics. A larger-than-reported population would prevent accurate urban planning, while straining sanitation capacities, the transportation sector, and access to resources such as water. The problem is even worse in rural areas. With an estimated 400 million to 500 million surplus agricultural workers among the 900 million rural residents,[98] farmers must eke out a living on tiny plots (one-sixth of an acre per person in some areas).[99] As a result, in 2000 fully one-eighth of China's population remained below the World Bank's absolute poverty line of $1 a day.[100] Farmers are forced to farm such small plots intensively, further degrading the land and contributing to soil erosion, flooding, and desertification. The consequences for both the environment and for the welfare of the rural population if such an imbalance between population and resources continues or worse, increases, will be dire.[101]

The Quest for Security

Mao's initial opening to the West toward the end of his tenure and the significant outward economic orientation of the reform leaders for more than twenty years have dramatically changed the calculus of external threat within China. During the Imperial era, and most of the Mao years, the quest for state security was a central contributing factor in the rampant degradation of the environment. War preparedness has not, however, been a significant factor contributing to environmental degradation during the reform period. Yet, some in the leadership continue to fear that tighter integration with the international community will breed dependency and contribute to a loss of sovereignty that could be dangerous should the outside world turn hostile. The mid-1990s quest for grain self-sufficiency reflects this latent security concern and illustrates how security concerns may continue to harm the environment.

In 1994, American environmentalist Lester Brown published a controversial article "Who Will Feed China?" in which he predicted that a declining capacity for grain production coupled with a surging demand for food in China would produce not only soaring levels of grain imports in China but also dramatic increases in the price of food worldwide. During the mid-1990s, Chinese grain imports did soar.[102] The piece elicited an unexpected political firestorm in China, arousing dormant but deeply ingrained beliefs about the centrality of grain in ensuring China's self-sufficiency and security. In early 1995, Xie Zhenhua, director of the National Environmental Protection Agency flatly stated, "Who will feed China? The Chinese people will feed themselves."[103] Speech after speech by Chinese leaders decried the country's growing reliance on international grain markets. President Jiang Zemin criticized the coastal areas for their large purchases of grain from the world market[104] and their "drastic drop in the amount of acreage under cultivation and decreasing yields."[105] An editorial in China's most prominent Party newspaper the *People's Daily* (*Renmin Ribao*) also attacked local officials for evading their responsibilities in grain production: "Some comrades believe that seizing the opportunity to accelerate development at present means undertaking a business which yields quick profit. There are also some comrades who unrealistically place their hope on other regions and rely on the state to solve the grain problem. They think that grain can always be bought with money."[106]

The response was a series of government-sponsored policies designed to increase Chinese grain production. Indeed, Chinese grain production during the successive four years swelled in response to these initiatives as well as favorable weather. China's reported grain yields rose dramatically from 394 million tons in 1994 to 508 million tons in 1999,[107] making it consistently the largest producer of grain in the world. In 2000, however, grain production declined by 9 percent as a result of the worst drought in fifty years.[108] And in 2001, drought continued, destroying millions of hectares of farmland and killing hundreds of thousands of livestock,[109] resulting in a further drop in grain yields of 2.1 percent and a total output of just over 452 million tons. In 2003, grain output dropped to a ten-year low of 435 million tons.[110]

But the central government's policies only exacerbated already troublesome trends in agriculture and land-use practices. First, the government's urgings did little to persuade the fertile but already economically dynamic coastal provinces to shift from industrial to agricultural development. They perceived little economic incentive to pursue agriculture. According to Jiangxi Director of Agriculture Liu Chuxin,

> It is now the universal view in all localities that they see slow returns from agricultural investment or no returns within a short time. They argue that no one will starve to death if less investment is made in agriculture. Therefore, given the repeated calls for increasing investment in agriculture and the great pressure from the authorities, some departments will play tricks with figures, creating a false image. They have increased investment in figures. But in fact, they have done nothing at all.[111]

Given the opportunities that rural industry presented for far more rapid growth, it is not surprising that some officials also perceived a focus on agriculture as a sign of a less developed region. In the village of Xishan in Shandong Province, once a productive farming community, the head of the village stated, "Not even a single villager grows grain now. We're not country bumpkins here."[112]

Recognizing the difficulties in reversing the trend toward industrialization in the wealthier provinces, in 1994, Beijing attempted to offset the loss by expanding the amount of cropland in other provinces by requiring that "all cropland used for construction be offset by land reclaimed elsewhere."[113] The unfortunate result was to encourage coastal provinces such as Guangdong and Jiangsu, which were losing extensive amounts of cropland to development, to pay other provinces such as Inner Mongolia, Gansu, Qinghai, Ningxia, and Xinjiang to plow land to offset their losses.[114] These provinces then reclaimed vast tracts of grasslands and continued to plow more land, which exacerbated soil erosion, water shortages, and desertification.

A U.S. Intelligence study by MEDEA, a group of renowned social, physical, and natural scientists, summed up the environmental situation:

> China's land productivity potential is subject to environmental limitations. There is little potential for increasing production by bringing additional lands under cultivation. Loss of land due to industrialization and urbanization, and loss of fertility due to erosion, salinization, effects of air pollution, and fluctuations in weather patterns are limiting factors. . . . A water balance model of the five "breadbasket" river basins in North China showed that two of them are currently operating at a deficit.[115]

Thus, the campaign not only failed to achieve the desired results but also produced a system of perverse incentives that encouraged Chinese officials and farmers to worsen the situation. Moreover, the campaign in no way advanced the use of technologies or practices that would actually increase grain yields. A telling report produced by Ma Zhong, a leading environmental economist in China, illuminates the results of China's focus on high grain yields rather than sound farming, including soil erosion resulting in a loss of five billion tons of topsoil per year, desertification, and pollution from chemical pesticides, which has affected 20 percent of China's farmland.[116] Overall, the quality of cultivated land in China is low; about 30 percent suffers from soil erosion, only 40 percent has access to irrigation, and 80 percent is classified as medium and low productivity.[117]

In Heilongjiang Province, the country's largest grain production base, accounting for 6 percent of the country's GDP, the situation is dire. The black soil, the most fertile soil on earth, is eroding; it is now only 20–40 centimeters (cm) thick compared with 40–100 cm fifty years ago, and "15 percent has been washed away entirely, leaving the barren yellow soil in sight."[118] To counter these trends, Ma encourages the expansion of "traditional" eco-farming, in which farmers rely on organic fertilizers, crop rotation, and cultivation of the natural enemies of insect pests and weeds.[119] Such practices currently are followed on only 3 percent of China's cropland.

While national security concerns no longer play a leading role as a source of environmental degradation in China, they remain latent in the consciousness of many within China's leadership and may emerge at times of international or domestic tension. The recent economic development campaign to "Open the West," which is explored in chapter 7, for example, has made explicit China's border concerns with Tibet and Xinjiang and raises difficult policy choices with regard to balancing rapid economic development and protecting the region's natural resources.

The Broader Costs

As China's leaders continue to press forward with economic reform, they must contend with a number of secondary, complex social and economic challenges engendered by the interplay of reform and the environment. Migration from rural to urban areas and the ensuing social tensions; growing instability in rural and urban areas; increasing public health problems; and significant economic costs from environmental pollution and degradation in terms of lost days of work, factory shutdowns, and costly pollution remediation programs all threaten to derail or slow China's economic growth and threaten the nation's social stability.

Migration

The migration of farmers and rural laborers in search of greater economic opportunity has been an integral part of China's socioeconomic landscape for centuries. While Mao maintained strict residence requirements for all Chinese, the reform period granted rural workers far greater freedom to move in search of a better life. As economic reform geared up, migrants became an essential element of the service workforce of many large cities, working in areas such as garbage collection and construction. Throughout the country, the migrant worker population is estimated at 120 million; in major cities, this "floating population" is estimated to constitute 10 percent to 33 percent of the population.[120] In Beijing, for example, it is estimated that one out of every three people is a migrant.[121] In Jiangxi Province alone, the outflow of farmers increased from 200,000 in 1990 to more than 3,000,000 in 1993. Chinese officials in such poorer inland provinces explicitly encourage migration to the wealthier coastal regions because these workers remit substantial sums of money back to their families. On average, the migrant population earns ten to twenty times as much in urban employment as they did in farming.

While thus far migrant workers have integrated with relative ease into the burgeoning economies of cities, the future is much less certain. Increasingly, agricultural workers are moving not because of the economic opportunity in the cities but because of environmental degradation in the countryside. In June 1991, *People's Daily* reported that during 1983–1990, 320,000 farmers from Gansu and southern Ningxia had to be relocated from a desertifying, eroded, and overpopulated mountainous region to newly irrigated areas along the Yellow River. This number was expected to reach 900,000 by 2000.[122] Today, a similar, though far smaller, relocation of people outside Beijing due to desertification is occurring. These are only small samples of a much larger problem. Chinese and Western analyses both suggest that during the 1990s, 20–30 million farmers were displaced by environmental degradation and that at least 30–40 million more may be relocated by 2025.[123] While much of this relocation may be accomplished peacefully, farmers, if not adequately consulted and properly compensated, will resist being moved as they have during the resettlement process for the Three Gorges Dam, straining the resources and popular support of the government.

In an effort to manage population flows and realize the potential benefits of urbanization, the government is in the midst of organizing a massive program to build new cities. These cities will afford their residents access to running water and sanitation systems, among other environmental and social benefits. One estimate is that 300 million farmers will be resettled, some forcibly, during 2002–2007.[124] On the positive side, this effort will remove people from degraded lands and deforested regions. However, it will also result in the loss of millions of acres of arable land to urban development. Moreover, in the short term, many are concerned that vast urbanization and migration will seriously damage the environmental quality of the cities. According to one study, migrants already account for 60 million tons of Shanghai's 110 million ton annual water shortage. The press report that half of Shanghai's building sites do not meet cleanliness standards, and the city has serious problems with its sanitation system: "When it comes to sewage treatment, much of the city of 14 million has been living in the pre-industrial age."[125] Some officials fear that planned new urban areas similarly will not develop adequate sanitation systems.[126]

The challenge of incorporating migrants into the social fabric of their new homes certainly is not limited to their impact on the environment. Stability in urban areas is increasingly in doubt as growing numbers of state-owned enterprise workers are laid off. Local economies are unlikely to be able to absorb the potential tens of millions of unemployed workers from newly merged or closed down SOEs.

As tensions have flared in urban areas over firings and growing unemployment, the government has attempted to discourage migration to the cities. In rural areas, the government has lowered water prices—in some cases to 80 percent lower than in Beijing[127]—to "keep farmers in their fields in order to avoid the huge social implications of a rush on the cities."[128] Meanwhile, some cities have moved to discourage migration. In June 2001, for example, officials in Changchun, the capital of Jilin Province, raised the administrative fees for pedicab drivers—almost all laid-off workers and farmers from the surrounding countryside who had fled drought-plagued farmland—to encourage their return to the countryside. The officials demanded a fee of 2,800 yuan (U.S. $300), or roughly five months earnings, for a license. The drivers protested, blocking the entrance to a local government compound.[129] Despite the incentives and sanctions, the migration continues.

If not managed properly, the flood of migrant laborers and unemployed workers could trigger serious conflict in urban areas. Already, urban residents blame these migrants for declining living standards, growing pollution, and rising crime rates.[130] According to Hong Kong's Information Center for Human Rights and Democracy, the number of large-scale protests in China has nearly tripled to 170,000 in 2000 since 1998.[131] Many of these demonstrations are in cities where unemployed SOE workers and migrant laborers make for a potentially explosive combination.[132]

Public Health

For Chinese citizens, perhaps the most frightening consequence of environmental pollution has been the range of public health crises popping up throughout the country. Although a system of country-wide environment-related health statistics has not yet developed in China, reports of specific, localized disease related to pollution are increasingly available, and are evident throughout the country:

- The devastating outbreak of Severe Acute Respiratory Syndrome (SARS) in China in 2003 underscored not only the embattled state of China's public health system but also the contribution of environmental factors to the severity of the problem, including improper disposal of medical waste and the potentially heightened risk of death from SARS in highly polluted cities.[133]

- In July 2000, the *Chinese Preventive Medicine Journal* published a survey indicating that in the wealthy region of Zhejiang, people whose drinking water had microcystin toxins were five to eight times more likely to die from intestinal cancer than those with access to cleaner drinking water.[134]

- In Binzhou village, Shangdong Province, the people suffer from brittle and cracking bones because of the impurities in the water they drink.[135]

- In the villages surrounding the famous wetlands of Baiyangdian, 70 miles south of Beijing, the water became so polluted that by the mid-1990s, liver and esophageal cancer rates in the region were three times higher than in areas with cleaner water.[136]

- Even in the suburbs immediately surrounding Beijing, the rice produced now evidences high levels of mercury.[137]

- Zinc mines in southern China have reportedly contaminated rice and shellfish with cadmium, contributing to high rates of anemia and kidney and bone disorders.[138]

Pesticides, too, have become a significant health risk. A report by China's Academy of Agricultural Sciences indicated that pesticide poisoning affects between 53,311 to more than 123,000 people annually.[139]

China's poor air quality, too, has serious implications for public health.[140] The World Bank has also estimated that 300,000 people in China die prematurely from air pollution annually, which is more than twice the number for South Asia, which has a roughly comparable population.[141] In the country's most polluted cities, when children breathe, it is the equivalent of smoking two packs of cigarettes per day.[142] China's long-time reliance on leaded fuel[143] has also contributed to lead poisoning among children in several major urban areas. Surveys undertaken in 2001 in Shenzhen and Beijing and in 2000 in Shanghai and Guangzhou indicated that more than 60 percent of children in Shenzhen, 20 percent of children in Beijing,[144] 50 percent of children in Shanghai, and 80 percent of children in Guangzhou were suffering from lead concentrations that exceeded the level considered safe by the WHO.[145] Lead poisoning can lead to lower IQ levels and other behavioral problems. Further, the life expectancy of traffic police in Beijing has been calculated to be roughly 40 years.[146]

For the government, the import of this public health challenge is magnified by the role that the problem plays in engendering social unrest and in the cost to the Chinese economy.

Environmental Unrest

In 1996, residents of Tangshan city took to the streets to protest the pollution emanating from Tongda Rubber, a tire-recycling plant. The air pollution from the factory was causing headaches, dizziness, nausea, rashes and insomnia.[147] In the face of their complaints, local officials threatened the protesters with the loss of their jobs and pensions. In a showdown, 700 residents blockaded the factory, while factory workers rallied against the plant closure. The local Tangshang government eventually ordered the plant shut down.[148]

More recently, in October 2001, hundreds of farmers demonstrated against Dadong Industries in Kunming, Yunnan for poisoning their crops with arsenic and fluorine. No one would buy the farmers' grain. After the protest, local officials purchased the grain from the farmers and urged Dadong Industries to control their pollution. But although Dadong Industries possesses pollution control equipment, it refuses to use it consistently.[149]

Damage to public health is one of the chief sources of environmental protests in China, many of which may turn violent. As anthropologist Jun Jing has noted in his review of 278 rural-based environmental disputes between the mid-1970s and the mid 1990s, 47 involved forceful popular protests such as sabotage and even riots.[150]

Resource scarcity is also a source of conflict in rural areas. In July 2000, about 1000 villagers in Anqiu, Shandong province, fought for two days when the police attempted to block peasants' access to makeshift culverts that were irrigating their crops. One policeman died, 100 people were injured, and 20 were detained.[151]

On rare occasions, such protests can trigger significant change in local environmental practices. Even so, the roots of the problem may not be addressed. In one case detailed by Jun Jing, in Dachuan village, Gansu Province, a fertilizer factory discharged its waste water into a stream that runs through Dachuan's fields before entering the Yellow River, which was until 1981 the village's only source of drinking water. Indeed, the factory was releasing not only chemicals but also incompletely burned fuel that left the water surface black much of the time. The local county government's environmental protection bureau did not take any action because the factory was a provincial level enterprise, whose jurisdiction extended beyond that of the county government. The drinking water was clearly polluted—a horse and thirty sheep went blind from the water; crops were damaged from excess ammonia; and many blamed birth defects and stillbirths on the drinking water. Still, the factory refused to take action to clean up the contamination.

In 1996, two hundred villagers, led by Dachuan's village head and a fish merchant, called upon the factory's Party secretary, general manager, and their families, demanding that they drink the water from the stream. If they were willing to do this, the villagers would bother the factory no more. When the management refused, several villagers drove up to the factory in tractors and, using rubber pipes, shot the contaminated water from the stream over the factory wall. After ten days, the factory agreed to repair a pump to provide tap water to more than 600 people in the village who were still drinking from the polluted river.[152]

When Jun returned to the village five years later, he found that the villagers had begun to transform the village into a replica of a Confucian village and a center for eco-tourism. The village is noted for its scenic beauty, encompassing mountains, wetlands, and rice paddies, and is home to a number of rare plants and birds. Noting that nearby city dwellers had long come to the village to fish in the more than 100 ponds, the villagers realized that for the city dwellers, it was less about catching the fish to eat than the opportunity to "come to nature."[153]

Still, deeper environmental values remain to be internalized. The fertilizer plant continues to operate without waste treatment facilities; the villagers have merely adapted the situation to their needs, building a pipe that will carry the factory's wastewater through the village and dump it downstream. Moreover, during a court battle with another village over control of an island in the middle of a reservoir that separated the two villages, residents of both villages chopped down nearly century-old trees as a means of ensuring that neither side would gain sole benefit from the timber.[154]

While the Dachuan situation resolved itself in a peaceful, if not entirely environmentally sound, manner, others do not. Farmers are increasingly squeezed by the demands of corrupt local officials who levy excess fees, a slowing rural economy, and declining yields on increasingly degraded land. In response, they are protesting in large numbers.

Beijing officials are clearly aware of the growing threat of rural instability, and the role of environmental degradation as a source of such instability. The Central Committee's organization department released a 308–page report that detailed the alarming increase in confrontational protests across the board, noting, "Protesters frequently seal off bridges and block roads, storm party and government offices, coercing party committees and government and there are even criminal acts such as attacking, trashing, looting and arson."[155] And China's minister of public security stated openly,

"Incidents [that] broke out over disputes over forests, grasslands, and mineral resources" are among "four factors in social instability."[156] While many reports suggest that violence over environmental issues is primarily based in rural areas, according to Jun's study, urban dwellers may be more willing than rural residents to engage in "blockades, sabotage, and even collective violence."[157]

Economic Costs of Environmental Degradation

As local officials confront the social costs of environmentally degrading behavior, they must also negotiate the skyrocketing economic costs. There is widespread agreement among environmental economists that the total cost to the Chinese economy of environmental degradation and resource scarcity is between 8 percent and 12 percent of GDP annually.[158] For example, the World Bank in 1997 estimated that total air and water pollution costs were $54 billion annually, or about 8 percent of GDP. When combined with resource shortages, such as water scarcity, some experts increase the costs of environmental degradation to 12 percent of GDP. In many industries and localities throughout China, continued economic growth is stymied by the environmental consequences of decades of unfettered economic development.

Health and productivity losses associated with urban air pollution top the list of economic costs: "hospital and emergency room visits, lost work days, and the debilitating effects of chronic bronchitis, are estimated at more than U.S.$20 billion."[159] Water scarcity in Chinese cities further costs about $14 billion in lost industrial output (when factories are forced to shut down), and in rural areas, water scarcity and pollution contribute to crop loss of more than $5 billion annually.[160] In 1999–2000, pollution-related problems reduced the country's fish output by 500,000 tons. While seemingly a small shortfall in an overall fish and seafood output of almost 43 million tons, the loss nevertheless cost roughly $370 million.[161]

Even more striking, however, are the growing costs incurred from expensive river diversion plans to meet growing water shortages in the north and west. Even as cities grapple with the social implications of rising water prices and the necessity of improving water efficiency and pollution prevention, they increasingly look to river diversions as a costly but less politically volatile alternative. Currently moving forward, for example, is the $58 billion South-North Water Transfer Project that will bring water from the Yangtze River to Beijing and Tianjin from provinces further south such as Hubei.[162] Dalian is also planning a river diversion project with an estimated price tag of $120 million.[163] Such projects bring with them not only substantial expenses but also the threat of intrastate conflict and protests by displaced citizens.

Looming Crisis

The task of protecting the environment in the reform era presents a number of complex political and economic challenges. For example, deforestation contributes to biodiversity loss, soil erosion, and flooding. Yet with pressures high to maintain employment in both rural and urban areas, protecting forested land puts millions of loggers out of work. Efforts to maintain high grain yields lead to overplowing of already degraded lands and a growing threat of desertification, while valuable fertile land is sold off at below market prices to industry, claimed for infrastructure, or incorporated into urbanization priorities. As economic opportunities in the countryside diminish, moreover, migration becomes an increasingly attractive option for millions of farmers; yet economic and environmental stresses in urban areas sharply constrain the willingness of officials and urban residents to accommodate the influx of migrant farmers.

Unless significant steps are taken, environmental trends suggest that resource degradation and scarcity will only continue to grow, as will the cost to the Chinese economy. One estimate predicts a 25 percent loss in arable land, a 40 percent increase in water needs, a 230 percent to 290 percent increase in wastewater, a 40 percent increase in particulate emissions, and a 150 percent increase in sulfur dioxide emissions by 2020.[164] Although little systematic work has estimated the future costs of these growing environmental threats, the World Bank has predicted that unless aggressive action is taken, the health costs of exposure to particulates alone will triple to $98 billion by the year 2020, with the costs of other environmental threats similarly rising. Moreover, the impact of environmental degradation and pollution extends far beyond the costs to public health and the economy to impinge on core leadership concerns over stability in the countryside and urban areas.

To date, China's environmental protection efforts have been dwarfed by the magnitude and complexity of the environmental challenge in the country. Throughout most of the country, an effective response is further hampered by China's bureaucratically weak environmental protection apparatus, as we'll explore in the next chapter.

CHAPTER FOUR

THE CHALLENGE OF GREENING CHINA

Introduction

Over the past two and a half decades, the rate of environmental pollution and degradation in China has far outpaced the capacity of the state to protect the environment. But this does not mean that China's leaders have done nothing. To the contrary, they have moved aggressively during the reform period to establish formal institutions, draft laws, and undertake large-scale programs in the name of environmental protection.

In many respects, their environmental strategy resembles their economic strategy. China's leaders provide administrative and legal guidance but devolve far greater authority to provincial and local officials; they utilize campaigns to implement large-scale initiatives of nationwide importance; they embrace the market as a force for change; and, as we'll explore in chapters 5 and 6, they rely increasingly on private citizen initiative and the international community to provide critical financial and intellectual capital.

Yet this mix of reforms, while wildly successful in spurring economic development, has proved insufficient to meet the challenge of protecting the environment. Without a strong central apparatus to serve as advocate, monitor, and enforcer of environmental protection, other interests generally prevail. Local officials, confronted with a choice between upholding environmental protection laws and supporting a polluting factory employing thousands of local residents, for example, usually choose the latter, considering environmental protection a costly drag on their local economy. Moreover, both local environmental protection officials and local judicial authorities are beholden to the local governments for their funding. Unsurprisingly, conflicts of interest frequently are resolved in favor of local officials' priority on economic development.

Economic reform campaigns typically, although not always, relax central control over the economy, providing significant incentive to many actors to fulfill the campaign's goals. In the environmental arena, however, there are few incentives, economic or otherwise, for local officials to carry out the initiatives of the central government. When market-based approaches to environmental protection, popular in Europe and the United States, are implemented in China, they often lack the necessary administrative, market, and enforcement mechanisms. China's weak legal tradition, too, enables corruption to flourish, although there are increasing opportunities for Chinese citizens to seek redress for environmental wrongdoing through the judicial system.

Despite all these obstacles, China's leaders over the past twenty-five years have achieved some success in establishing institutions and norms to protect the environment and, perhaps more important, have laid the foundation for a transformation in how environmental protection will be integrated into future economic development. Whether this potential is achieved, however, will depend on the willingness of China's next generation of leaders, beginning with Chinese Communist Party general secretary and president Hu Jintao, to build aggressively on the initial steps taken under Deng Xiaoping and Jiang Zemin to strengthen significantly the central institutions responsible for environmental protection and enhance the necessary adjunct institutions such as the judiciary and banks.

First Steps Toward Environmental Governance

Even before the ascension of Deng Xiaoping in 1978, the winds of environmental protection in China had begun to shift. In 1972, three events sparked a new environmental consciousness in Beijing.

The first two were ecological disasters. In the northeastern coastal city of Dalian, the beach turned black, millions of pounds of fish were lost, the port became clogged from polluted shells, and dikes eroded. Other coastal cities recorded similar incidents. That same year, tainted fish from the Guanting Reservoir outside Beijing appeared in the city's market. These events prompted then premier Zhou Enlai to establish a small leading group composed of officials from some of the country's provinces and largest cities to address the issue of treating and protecting the water resources of reservoirs.[1]

Perhaps most significant, however, the year 1972 witnessed the United Nations' (UN) first international environmental conference, the UN Conference on the Human Environment (UNCHE). This conference planted the seeds of environmental change in China and signaled a turning point in China's approach to environmental governance. In keeping with China's decision to reestablish political and economic relations with the rest of the world and the country's assumption of the Chinese seat in the UN in 1972, Premier Zhou Enlai sent a delegation to the UNCHE in Stockholm, Sweden,[2] opening the door to a new understanding of China's environmental problems and potential solutions.

While at the UNCHE, the delegation shied away from an open discussion of the country's environmental practices and experiences and sought to portray the issue of environmental degradation and protection in the context of conflict between the developing world and the superpowers. The official statements of the Chinese delegation largely reflected cold war rhetoric:

> The Chinese delegation is opposed to certain major powers practicing control and plunder under the name of the human environment, and the shifting by these Powers of the cost of environment

protection onto the shoulders of the developing countries under the guise of international trade. . . . The urgent task before the developing countries is to shake off the plunder undertaken by imperialism, colonialism, and neo-colonialism of various descriptions, and to develop their national economies independently.[3]

The delegation also proposed a set of ten principles to be incorporated into the final declaration of the conference, which included demands to:

- assure the right of developing countries to develop first and address their environmental challenges one by one;
- reject the "groundless" nature of others' "pessimistic view" in respect to the relationship between population growth and environmental protection;
- ban biochemical weapons and prohibit and destroy all nuclear weapons;
- assign responsibility to the superpowers for the destruction of the human environment through their "imperialist" policies of plunder, aggression, and war;
- sanction countries that plundered and destroyed the environment of developing countries;
- fight collectively against pollution;
- compensate any country polluted by another;
- support the free transfer of scientific and technical knowledge;
- develop an international fund by the industrial countries to support environmental protection elsewhere;
- ensure a country's sovereignty over its natural resources.[4]

In the end, some of these principles were incorporated into the UNCHE declaration, including the need to fight collectively against pollution and the importance of the free transfer of scientific and technical knowledge. Others were directly refuted, including China's assertion that no relationship existed between population growth and the environment. Most were simply not adopted.[5]

Upon the delegation's return to Beijing, its report prompted Zhou

Enlai to take several steps toward establishing a national environmental protection apparatus. In June 1973, Premier Zhou organized the country's first National Conference on Environmental Protection. One year later, in May 1974, the State Council established a top level interministerial Environmental Protection Leading Group of the State Council to study environmental protection issues, although it only met twice during the following nine years.[6] In addition, all provinces, municipalities, and autonomous regions were required to establish organizations for environmental control, research, and monitoring. Local governments also formed "three wastes offices," which had little bureaucratic authority, but nonetheless represented the first foray into institutionalized environmental protection at the local level. Through the mid-1970s, other small steps were taken to improve the environmental situation, including some environmental investigations by local officials. During this period, however, the political turmoil and large-scale development campaigns of the Cultural Revolution prevented real progress on environmental protection.[7]

As they did in the realm of the economy, Deng Xiaoping and his supporters greatly accelerated the development of a nationwide environmental protection effort. China's leaders recognized that China's path of economic development was having a ruinous effect on the environment, as noted in a report published by the Environmental Protection Leading Group:

> Environmental pollution is spreading in our country to such a serious degree in some areas that people's work, study and life have been affected. It also jeopardizes people's heath and industrial and agricultural development. . . . It is an important component of our socialist construction and modernization to eliminate pollution and protect the environment. . . . We should not follow a zigzag path of construction first, control second. Environmental problems should be dealt with during construction.[8]

In 1978, the Chinese constitution was amended to acknowledge concern for the environment, with the proviso that the state had to protect the environment and natural resources as well as prevent pollution and other public hazards. And in 1979 the National People's Congress (NPC) approved the draft Law of the People's Republic of China on Environmental Protection, which established basic principles to safeguard the environment and promoted the development of a legal network for environmental protection.[9]

In the 1970s and early 1980s, the Chinese government held a series of important meetings and passed regulations to control industrial and marine pollution. In addition, it enacted several structural reorganizations designed to strengthen the environmental protection bureaucracy—although some in fact had the opposite effect. In 1982, for example, the government created a Ministry of Urban and Rural Construction and Environmental Protection that incorporated the Leading Group on Environmental Protection. Many local governments followed suit, merging their environmental protection bureaus (EPBs) into new departments of urban and rural construction, in the process eliminating jobs and diminishing the bureaus' already weak authority.

Two years later, in the wake of the second National Conference on Environmental Protection in January 1984, the State Council in effect acknowledged its mistake in abandoning the Leading Group and established an Environmental Protection Commission. This commission, composed of participants from more than thirty government ministries and bureaus, reviewed environmental policies, initiated new plans, and organized environmental activities, such as inspections, into local implementation of environmental laws.[10] Later that year, the State Council also raised the profile of the Environmental Protection Bureau, elevating its status to the National Environmental Protection Bureau within the Ministry of Rural Construction and Environmental Protection, doubling its staff to 120 people, and enabling the agency to "issue orders directly to provincial EPBs, decide upon and conduct its own meetings, and receive Ministry of Finance funds directly earmarked for environmental protection, rather than waiting for these funds to be channeled through the Ministry of Construction."[11]

The following year, the Chinese leadership appointed Qu Geping the first chief administrator of the National Environmental Protection Bureau. Qu was an inspired choice to lead China's still-nascent environmental protection effort. With a background in chemistry, Qu had spent his early career at the Ministry of Chemical Industry and the State Planning Commission. A protégé of Zhou Enlai, he had participated in the 1972 UNCHE and served as China's representative to the UN Environment Programme during 1976–1977. While scholarly and possessed of a gentle sense of humor, Qu was also a skilled politician, respected by Chinese and foreigners alike for his willingness to speak frankly concerning China's environmental challenges.

Qu worked relentlessly to create an independent environmental agency; four years later, in March 1988, his efforts bore fruit when the National Environmental Protection Bureau, also referred to as the National Environmental Protection Agency (NEPA), finally achieved independent status from the Ministry of Urban and Rural Construction and was able to report directly to the State Council (although its bureaucratic standing remained one notch below that of a ministry).[12] In 1989, the standing committee of the NPC formally promulgated the Environmental Protection Law, which embraced four central principles: (1) coordination of environmental protection, (2) pollution prevention, (3) polluter responsibility, and (4) enhancement of environmental management.[13]

As China entered the 1990s, guarded optimism about the future of environmental protection prevailed. Although officials at the Third National Conference on Environmental Protection in May 1989 acknowledged some serious failings in the implementation of environmental regulations, they remained hopeful. Qu commented that the government now realized that environmental problems could exert a significant impact on development, on the strength of the country, and on the stability of the society.[14]

Much of the next decade, however, would suggest that China's leaders were not yet ready to bridge the gap between recognizing the importance of protecting the environment and acting to respond to the challenges. The reality of environmental protection remained much the same as it had at the end of the 1980s. As in many countries, China's environmental protection agency had little clout in interministerial negotiations. Few local officials paid attention to environmental protection laws, secure in the knowledge that environmental protection was not a central priority, and focused instead on raising the economic standards of their local citizens. Moreover, the imbalance between the rate of environmental degradation and pollution and the country's capacity to respond was both understood (however imperfectly) and accepted by many. "First development, then environment" remained a frequently articulated principle both within the Chinese leadership and more broadly in the country's media from the early 1980s to the mid-1990s.

Opening the Door: Rio and Its Aftermath

The two decades from the 1972 UNCHE to the 1992 UN Conference on Environment and Development (UNCED) in Rio de Janeiro, Brazil, were a period of profound economic, political, and social change in China. Yet in the months leading up to the UNCED there was little in the Chinese approach to the conference that evidenced such change. Prior to the UNCED, Chinese negotiators articulated five principles of environmental protection that were essentially an abridged version of those its delegation had delivered twenty years earlier at the UNCHE:

- environmental protection can only be effective when development has been attained;
- the developed countries are responsible for global environmental degradation;
- China should not talk about its responsibility for global environmental pollution and degradation;
- the developed countries should compensate developing countries for the efforts they undertake to meet international environmental agreements and should provide environmental technology and intellectual property at below-market prices;
- and the sovereignty of natural resource rights must be respected.[15]

To many international observers at the Rio conference, China was an inflexible obstructionist, intent on allying the developing countries against the advanced industrialized nations to prevent an international agreement on climate change, one of the key topics of the gathering.[16]

As was the case twenty years earlier, however, China's recalcitrance on the international stage was followed by a major shift in the consciousness and politics of the environment on the domestic front. The Chinese who prepared for and participated in the UNCED developed a new vocabulary for environmental protection. Chinese officials began to incorporate the ideal of *sustainable development*[17] into their planning process; in some cases, this was merely rhetorical, but in others the language represented real policy change, as discussed below.[18]

The UNCED also triggered a profound change in the Chinese leadership's conception of environmental governance by highlighting Western ideals of popular participation and reintroducing the notion of the nongovernmental organization (NGO) into Chinese society. While the centerpiece of the conference was the formal negotiations by the member countries, a parallel meeting of NGOs from the participating countries received even greater international attention. China's participation in the NGO forum, however, was limited by its inability to deliver any genuine NGOs; instead, China was represented by a set of government-organized nongovernmental organizations (GONGOs). This was a source of embarrassment to some senior leaders. Thus, one year later, in June 1993, the government for the first time cited public participation as a goal in environmental protection.[19] Around the same time, when a well-known Chinese historian, Liang Congjie—urged on by university students and friends, including a NEPA vice director—proposed the establishment of China's first environmental NGO, his proposal was welcomed by some government officials. Although governmental provisions for NGOs existed in China prior to this time, no formal application had been made to establish one. So Liang's effort represented a breakthrough for China.

The UNCED also spurred NEPA to issue an annual "Communique on the State of China's Environment" which reported on the state of the environment, progress made, and goals outstanding.[20] For Chinese officials not formally involved in environmental protection, the UNCED enhanced their awareness of the value of international contacts and assistance beyond bilateral and international governmental organization (IGO) linkages.[21]

The Rio conference also reinforced notions already being advanced by the World Bank and other international actors about using the market as a basis for environmental policy reform, through measures such as raising the price of natural resources to properly reflect their economic value. Market-based pricing for coal was one of the first such reforms; in the mid-1990s, the government increased and deregulated coal prices. In many regions, this meant that new, higher coal prices finally covered the costs of production and delivery.[22] However, the full environmental costs of mining and burning coal were never incorporated into energy pricing, making it far cheaper to continue to use coal than, for example, to exploit the country's as yet largely untapped natural gas reserves.

So China made genuine advances in environmental protection in the wake of both the UNCHE and the UNCED. Nonetheless, Chinese environmental protection leaders remain deeply concerned about the future. In a speech at Qinghua University in April 2001, Xie Zhenhua, Qu Geping's successor as head of NEPA (which became the State Environmental Protection Administration [SEPA] in 1998) painted a dark picture of the current state of the environment in the country, noting that "the amount of pollutant discharged into the environment far exceeded accepted levels . . . the government had not effectively put a stop to the deterioration of the eco-system . . . [and], although the government had made progress in protecting the environment during the past five years, the future remained grim."[23]

Challenges of China's Bureaucracy

Underlying Xie's unhappy future scenario is the continued weakness of the state's environmental protection efforts. The core agencies behind China's environmental protection efforts—the National People's Congress (NPC), which is the top lawmaking body in the country;[24] SEPA;[25] and the judiciary, headed by the Supreme People's Court—together claim responsibility for the full scope of central governmental activities, including drafting of laws, monitoring implementation of environmental regulations, and enforcement. Yet each element of this bureaucratic apparatus exhibits fundamental structural weaknesses that undermine the best of intentions.

The Legal System

China's legal system has been widely criticized for its lack of transparency, ill-defined laws, weak enforcement capacity, and poorly trained advocates and judiciary.[26] In 1993, however, China's leaders addressed at least some of the weakness in the lawmaking process, establishing specialized committees within the NPC[27] on issues such as agriculture and rural affairs, and internal and judicial affairs. For the environment, the Chinese leadership tapped Qu Geping, head of NEPA, to head the new Environmental Protection Committee, soon renamed the Environmental Protection and Natural Resources Committee (EPNRC)[28] within the NPC. Under Qu's leadership, the committee spearheaded an impressive environmental lawmaking effort. By 2001, China had passed 7 environmental protection laws and more than 123 administrative regulations.[29] (These laws are complemented by the more than 20 technical environmental regulations issued by the State Council, such as the "Implementing Regulations for the Water Pollution Prevention and Control Law," and the 100 environmental rules and methods and 350 standards formulated by SEPA and other State Council ministries and agencies.)

Qu retired in 2003, but he was relentless in his effort to improve the quality of China's laws. According to many environmental protection officials and experts in both China and the West, most Chinese environmental protection laws are too broad, providing local officials with little guidance on implementation. China legal expert William Alford has remarked that China's environmental laws are like policy statements rather than laws in the Western sense. For example, John Nagle illuminates the difference between China's water pollution law and that of the United States:

China's Water Pollution Prevention and Control Law states that "[e]nterprises and other undertakings which cause serious water pollution must eliminate pollution within a stipulated time." What constitutes "serious" water pollution is not self-evident, nor is the meaning of "enterprises and other undertakings." . . . The U.S. Clean Water Act, by contrast, specifies in much greater detail the type and amount of pollution that particular sources may emit through the procedures for establishing effluent limitations and water quality standards and through the permit process. Additionally, the Clean Water Act compels any point source to obtain a permit stating the exact amount of each substance that may be discharged into the water.[30]

It is therefore difficult for the Chinese to know what is prohibited and what can be called to account through legal redress. In one case, in Changzhou, Jiangsu Province, when a local environmental protection bureau attempted to sue a local chemical company for failing to pay its discharge fees over a two-year period, the bureau at first couldn't decide whether the basis for the suit rested on the enterprise's failure to pay four small amounts for its discharge fees or its failure to pay overstandard fees. Moreover, the Changzhou court, scheduled to hear the case, pointed out that inconsistencies between national regulations and Changzhou's rules might make a suit difficult.[31] Ambiguity in the laws also permits conscious exploitation by enterprises or other actors. As Xiaoying Ma and Leonard Ortolano point out in their study of environmental regulation, "The Chinese even have a common saying, 'national policies, local countermeasures' [*shang you zhengce xia you duice*], to describe the practice of exploiting the ambiguity of national laws and regulations to figure out ways around them."[32]

Furthermore, as in other countries, lawmakers must balance environmental protection needs against other social considerations. In China, the chief concern is the potential link between stiffer environmental protection laws and economic disruption. As Wei Haokai, a member of the forward-looking advisory body the Chinese People's Political Consultative Conference and scholar at the Chinese Academy of Social Sciences, argued in discussing the need for an environmental protection tax in 2001,

> [Despite the fact that] the government annually earmarks 50 billion yuan [$6 billion] for environmental protection, things have not markedly improved. Levying an environmental protection tax on enterprises is a remedy for the dilemma. Anyone culpable for environmental pollution should pay the tax. This is totally in sync with a market-oriented economy. . . . But how much environmental protection tax enterprises should pay must be carefully calculated. It should not become enterprises' burden. The tax should be charged according to how much damage an enterprise has caused or will cause to the environment. . . . Though levying an environmental protection tax on enterprises will, to some extent, increase enterprises' costs, blunt their competitive edge and lead to unemployment, it could stimulate the development of environmental protection enterprises and ensure the implementation of a sustainable strategy from the long term perspective. To reduce losses, the state should provide subsidies to deserving enterprises. Only when enterprises' worries are alleviated can this job be accomplished smoothly. But in the final analysis, environmental protection is teamwork. It requires the entire country to edge towards a consensus.[33]

Wei's analysis highlights the central tension inherent in China's environmental protection efforts: how to balance the development of the economy with environmental protection, while maintaining the government's belief that less central government involvement is better.

An equally challenging factor in China's laws is the exhaustive consultative process. Law drafting involves a wide range of ministry representatives, local officials, and technical experts, all of whom may lobby for significant changes until, frequently, the law is watered down to the point that it serves no real use.[34] The negotiations over the energy efficiency law, for example, lasted more than four years, and today the law is still not well implemented, a function of

both the complexity of the law and the weak enforcement apparatus.[35]

Yet China's environmental law makers have demonstrated increasing sophistication in their understanding of how to negotiate
and draft a technically sound and politically viable law. William Alford and Benjamin Liebman's fascinating study of the thirteen-year,
three-stage development of the Air Pollution Prevention and Control
Law demonstrates the learning process among China's environmental advocates. The first law, issued in 1987, was a "broad, but vague,
framework for the regulation of air pollution . . . consist[ing] of just
forty-one articles, filling less than eight pages of text in Chinese,
separated into six chapters."[36] By the mid-1990s, some of China's
environmental officials in the EPNRC and NEPA were convinced
that the law was not only "vague and unclear" but also rooted in the
assumption that the state would be playing a much larger role
through controlling air pollution generated by SOEs than it did, especially given the rise of TVEs and the transportation sector.[37]

The views of the EPNRC and NEPA were supported by some sympathetic municipalities such as Beijing, Shanghai, and Tianjin, as
well as by public opinion polls and the reality of China's worsening
air quality. The forces aligned against revising the air pollution law,
however, were formidable, including the State Planning Commission, the State Economic Commission, the coal and electricity ministries, provincial governments in Sichuan and Gansu (where the
coal produced and used was especially dirty), and the automobile industry. These groups argued that the economic and social costs of
undertaking pollution control measures such as washing coal or
stringent inspections on all new autos far exceeded the benefits.
Even the most critical element of the law—namely, changing the
method of measuring pollution from the concentration of pollutants
in emissions to total load[38]—produced a firestorm of opposition.
The compromise eventually achieved "eliminated the vast majority
of the EPNRC's key suggested revisions."[39]

By 2000, however, the EPNRC and NEPA's successor, SEPA,
spurred on by environmentally proactive municipalities, sought yet
again to revise the law. This time, however, their strategy was far

more inclusive at the outset, inviting views from industrial ministries, explaining their perspectives thoroughly, and incorporating studies of public attitudes toward the environment.[40] The law also took care to grant leeway to actors who might need more time to implement the more stringent air quality standards, and the NPC empowered the State Council to decide when individual cities would be required to meet air quality standards. The first cut targeted only 47 cities, since according to Qu Geping, "It is unrealistic to expect all the 668 cities in the country to reach the national air quality standard in a short time."[41] While it is too early to predict whether the latest version of the law will be successfully implemented, both in process and in outcome, it reflects a far more sophisticated understanding of the politics and economics necessary to improve the outlook for effective environmental protection. By the first half of 2003, of the 47 cities, only 18 enjoyed air quality that met the state-prescribed standards for more than 90 percent of the year, but 29 had registered improvement in air quality over the previous year.[42]

SEPA and Its Environmental Protection Bureaus

Management of environmental protection in China is shared by many agencies and other actors, depending on the issue. Resolution of large-scale water pollution problems, for example, might involve the Ministry of Water Resources, which is in charge of water allocation; the Ministry of Construction, which handles water and sewage treatment; the State Forestry Administration, which takes care of water within reservoirs in nature reserves; the State Oceanic Administration, which is responsible for coastal waters; and SEPA, which monitors water pollution in areas not overseen by other ministries. In addition, economic agencies such as the Ministry of Finance play a key role in determining the level of fines assessed factories for water pollution, and the Environmental Protection and Natural Resources Committee (EPNRC) of the NPC is responsible for coordinating and drafting laws regarding water pollution. However, involvement in the full range of environmental protection activities, including law drafting, monitoring, enforcement, environmental impact assessments, and research rests with the State Environmental Protection Administration (SEPA) and its local bureaus.

In many respects, SEPA's rise within the bureaucratic hierarchy of China's state apparatus mirrors the rise in the importance of environmental protection within the Chinese government. In 1998, as part of sweeping administrative reforms that reconstituted many institutions within the Chinese bureaucracy, then premier Zhu Rongji elevated the agency from its long-time subcabinet status to ministerial rank, changing the name from the National Environmental Protection Agency (*Guojia Huanjing Baohu Ju*) to the State Environmental Protection Administration (*Guojia Huanjing Baohu Zongju*). Many international observers considered this a significant step in China's commitment to environmental protection.

In addition, SEPA's enforcement capacity during the late 1990s began to be enhanced through others' efforts. The State Economic and Trade Commission (SETC), for example, took an active role in pushing for the implementation of cleaner production technologies. In 1999, SETC released the first list of outmoded production capacities, equipment, and products that it deemed should cease operation within a specified period.[43] SEPA, together with both the SETC and the Ministry of Supervision, undertook periodic large-scale central inspections of local environmental enforcement. On occasion, these inspections have engendered real change. One such inspection in 2001 led officials in Beijing and Chongqing to issue new environmental regulations, and other cities, such as Dalian, to establish pollution hotlines for the public.[44]

Despite such steps, a number of contributing factors have limited the broader influence of SEPA and other environmentally inclined actors. As noted, former premier Zhu Rongji—a reform-oriented leader in charge of China's economic development and considered a strong supporter of environmental protection within the Chinese leadership—elevated NEPA's status in 1998 as part of an overall reorganization of the state's bureaucracy. Yet shortly thereafter, during a lunch in New York with the newly named SEPA director, Xie Zhenhua, I asked Xie to describe the most important change for SEPA as a result of its new ministerial rank. He replied, with no small degree of irony, "They cut my staff in half." In fact, in 1998, all the ministries had taken sizeable losses in employees; moreover, some ministries had been dissolved and others, such as the Ministry of Forestry, had seen their rank dropped a notch. SEPA's already meager staff further decreased from 600 to 300 people.[45] Zhu simultaneously eliminated the overarching State Environmental Protection Commission, which was the only standing forum for top-level coordination on environmental protection. Such maneuverings have diminished SEPA's ability to coordinate high-level environmental policy and stretched its capacity very thin.[46]

SEPA's bureaucratic challenge is compounded by the low level of funding accorded environmental protection. Chinese investment in environmental protection had hovered around 0.8 percent of gross domestic product (GDP) during the ninth five-year plan (1996–2000), but increased significantly to 1.3 percent of GDP in the tenth five-year plan (2001–2005) (about 700 billion yuan or $85 billion).[47] This remains far below the 2.19 percent of GDP that some Chinese scientists believe is necessary simply to keep the environment from deteriorating further.[48]

In addition, as the environment has risen on both the domestic and international agenda, other bureaucratic actors have increasingly impinged on SEPA's mandate. The State Development and Planning Commission (SDPC), in particular, actively sought to expand its environment-related activities as its planning mandate has diminished over the course of the reforms.[49] The State Planning Commission and the State Science and Technology Commission,[50] for example, coordinated China's preparations for the UNCED even though the former lacked expertise in environmental matters. In the aftermath of the Rio conference, these two organizations became the chief conduits for the international funds that flowed to China for Agenda 21[51] projects through the newly created Administrative Center for China's Agenda 21, despite the concerns of the UN Development Programme, a major source of financial and technical aid to China, which advocated NEPA's inclusion in the Center's leadership. (NEPA responded by developing its own competing Trans-century Green Program to attract international support for its activities.) The SDPC (in March 2003, the SPDC, along with part of the SETC and part of the Structural Reform Office, became the State Development and Reform Commission) has also served as the chief broker in determining which environment-based projects are eligible to submit for World Bank financing, a role that has limited the number of projects, which have been disproportionately located in the wealthiest regions of the country.[52] In 2001, the SDPC became the lead agency for China's massive "Go West" campaign. Despite the prominent role given to "ecological construction" as one of the six major tenets of the campaign, SEPA was excluded from the twenty-two–agency leading group identified at the campaign's outset in 1999. According to some Chinese environmentalists, part of the difficulty is SEPA's traditionally low standing. At the same time, they note that Xie Zhenhua, while clearly committed to improving China's environment and SEPA's standing in the bureaucratic hierarchy, is not as politically adept as his predecessor, Qu Geping.

Environmental protection is even weaker at the local level. It rests with some 2,500 environmental protection bureaus (EPBs)[53] spread throughout the country, with roughly 60,000 employees; if offices at the township and village level that address environmental protection (along with other issues) are included, the numbers swell to 11,000 agencies and 142,000 personnel.[54] While nominally responsible to both SEPA and their local governments, EPBs rely on the latter for virtually all their support, including their budgets, career advancement, number of personnel, and resources such as cars, office buildings, and employee housing.[55] Not surprisingly, EPBs are typically quite responsive to the needs and concerns of the local government. SEPA maintains only a supervisory role with regard to the EPBs. Many EPBs are ill-equipped to manage the task of environmental protection at the grassroots level.

A report on environmental protection in Hebei, a province that ranks twelve out of thirty-one in per capita GDP, illuminates the challenge. In 1994, the Hebei provincial EPB reported that almost 90 percent of its rural environmental protection officials were unqualified, and their offices grossly understaffed. Of 139 counties in Hebei, only 17 had first- or second-rank environmental protection bodies (i.e., those legally empowered to enforce environmental protection laws). Therefore, only 12.3 percent of environmental protection offices at the county level could enforce the law independently. The other offices functioned as divisions of the counties' construction committee bureaus, planning committees, or urban construction bureaus. Twenty-one offices had no power to enforce the law.[56] Even as Hebei has increased its capacity with more qualified environmental protection officials over time, corruption has undermined some of its progress. In 2001, Xie Zhenhua personally intervened to sanction several EPB officials in Hebei for not adequately addressing the problems generated by a polluting factory.[57]

The administrative rank of the EPBs is also sometimes lower than that of the enterprises it is supposed to oversee. In one case, the Guangzhou Environmental Protection Office was not permitted to monitor the wastewater from the Guangzhou Paper Manufacturing Company because, administratively, the rank of the environmental protection office was lower than that of the company director.[58]

Limited resources and local politics have further contributed to reduce the efficacy of environmental protection efforts at the local level. Environmental protection bureaus are responsible for monitoring waste and pollution from enterprises and, if necessary, levying and collecting the fees. Although fee collection has increased over time,[59] barriers to effective monitoring and enforcement efforts remain.

First, monitoring and inspection teams are typically inadequate to the task. Even Shanghai, recognized as a model environmental city[60] under the leadership of former Mayor Xu Kuangdi, could not muster enough resources within its EPB to inspect the enterprises within its purview. In 1997, Shanghai organized a special team of 100 environmental inspectors. With more than 20,000 factories and a wide range of tasks for these inspectors, however, EPB officials feel overwhelmed, recognizing that they cannot visit every factory even once per year.[61] Shanxi Province, China's largest coal-producing region, has established a special environmental police corps, equipped with advanced transportation, communications, and monitoring capabilities, to monitor local industries, nature reserves, and construction projects to ensure adherence to the law.[62] Nonetheless, Taiyuan, the capital of the province, routinely ranks within the top five most polluted cities in China.

While most EPB officials claim they undertake surprise inspections, Guangzhou EPB officials admitted that they notify factories before they visit,[63] thereby ensuring that problems would likely go undetected. Even when inspections were a surprise, EPB officials uniformly acknowledge that factories were free to pollute at night and on days when there were no inspections. When central government teams from SEPA, the SDRC, the State Forestry Administration, and the Ministry of Supervision, or even provincial teams, undertake inspections sweeps throughout a province, they detect gross violations. In 2001, in Shandong Province, for example, provincial officials investigated 178 enterprises: 84 were exceeding legal emissions limits, and 11 previously ordered to close were still operating.[64]

When companies exceed pollution discharge limits for certain pollutants and by law must pay a pollution discharge fee, the process of collecting the fee is often plagued with difficulties. If an EPB official has a personal relationship with the head of an enterprise, a smaller fee is usually levied. In other cases, fees are not fully collected or properly disbursed according to government regulations,[65] which specify that as much as 80 percent will return to the enterprise to pay for improvements to its pollution control capacity and the remaining amount will support the environmental protection bureaucracy. In some cases, 20 percent of the fee goes to the environmental protection institutions, but the other 80 percent is forgotten. In addition, even if firms agree to pay the fee, they frequently argue over the amount or delay payment.[66] According to a study by the World Bank, social, political, and economic considerations often dictate which firms are more likely to be required to pay the full fee. Private firms and those whose pollution has a higher social impact (in particular if citizens complain) have less bargaining power with the local EPBs. State-owned enterprises and those firms with a precarious financial situation, however, have relatively greater bargaining power.[67] In some cases, even when firms pay the fee and receive the 80 percent to retrofit their factory, they use the money to pay for other needs of the enterprises' environmental protection departments rather than pollution prevention measures.[68]

The government has reported that it typically collects about 30 percent of the total funds actually owed; in 1996, this number reportedly reached approximately $500 million.[69] However, in several cities, including Shanghai, the recovery rate is much worse than 30 percent. In 1998, the Shanghai municipal and district environmental protection agencies recovered arrears of $1.8 million. This accounted for only about 12 percent of the money nominally charged for excessive pollution.[70]

Most startling, perhaps, the fee collection process has also produced a perverse incentive for EPBs to encourage the persistence of pollution problems. In Wuhan, for example, municipal EPB officials developed what they considered to be an effective mechanism to encourage local EPBs to enforce fee collection by appealing to local EPB officials' desire to increase their personal wealth. The Wuhan EPB was to set targets for fee collection; when these were exceeded, the extra money went into the pockets of the EPB officials. The EPB officials became financially dependent on the fee collection.[71] When factory managers complained about paying fees long after they had made the necessary changes, EPB officials simply claimed that they had no opportunity to do follow-up monitoring.[72]

In the most desperate of situations, the local EPBs are so underfunded that they use pollution discharge fees to pay their basic wages and administrative costs instead of putting the money toward environmental protection.[73] In 2001, a government investigation uncovered widespread abuse of discharge fees. An audit of forty-six cities found that almost $73 million was misappropriated for uses such as purchasing housing and automobiles or operating private businesses. All told, instead of devoting 80 percent of the discharge fee to pollution control, the EPBs devoted 44 percent.[74]

This corruption in the fee collection system apparently led the Chinese leadership to revise the process by which fees are collected; beginning in July 2003, the pollutant discharge fees were paid directly by the firms to a designated commercial bank, within one week of receiving a collection notice. However, this new set of administrative regulations continues to support the "reduction, exemption or postponement" of fees if a polluting firm is suffering serious economic loss or "facing some special difficulties."[75] Thus, the potential for arbitrariness and even abuse in the fee negotiation process remains.

The Judiciary

The third prong of China's environmental protection effort rests in its nascent system of advocacy. China historically has operated under the rule of men rather than the rule of law. During much of Mao Zedong's tenure, the law was perceived as a means of oppression, and qualified judges were dismissed in favor of party officials with strong ideological credentials. In 1978, however, Deng Xiaoping began to work to change this conception of the law and to develop a legal system modeled at least in part on western systems.[76] Unlike in the United States, however, past cases do not serve as binding precedents, although they are sometimes referred to for guidance.[77] The top judicial authority, the Supreme People's Court, is the only body permitted to issue official interpretations of the law.[78] As with the economic and environmental bureaucracies, the judiciary is also highly decentralized. The Supreme People's Court possesses a supervisory role but has no real power over the personnel or budgets of the lower courts. Such control rests with the local government.

Not surprisingly, this system raises problems for the independence of the courts. A study of the Chinese court system found that judges often reach decisions by "relying on CCP (Chinese Communist Party) policy, the views of the local government, and a court's individual sense of justice and fairness in contractual dealings."[79] Since courts are dependent on the local government for their funding, local officials wield a "potent weapon against judges who would be more willing to enforce the law despite adverse economic consequences."[80] As one environmental lawyer pointed out, "In all the suits that we have lost, the courts have not followed the law. Instead, they ignored the legal or technical merits of our case in order to support the local enterprises."[81] Even when a court order is issued, there is no guarantee that an EPB will be able to enforce its mandates.[82] Thus, when powerful local interests block enforcement of environmental protection laws, Party intervention is often necessitated. Such intervention, however, as Nagle notes, "robs the law of its independent force."[83] Chinese lawyers also complain that judges cannot render an informed ruling because they do not understand the basic facts of the case.[84] In many cases, judges are civil servants or demobilized soldiers with only a high school education and no training or practice in the law.[85] When one law professor questioned the logic of having military men with no legal training serve as judges, the People's Liberation Army (PLA) responded by pointing to the military's historical contribution to the country.[86] Nor is every lawyer or judge well trained. In the first eight months of 1998, Chinese officials estimated that nearly 10,000 cases were "incorrectly prosecuted or wrongly decided."[87]

Much of the problem for local EPBs in pursuing legal judgments against polluting firms is the lack of trained legal staff. While the number of environmental lawyers is on the rise, it remains small given the magnitude of the problem. Even in Shanghai, with one of the largest, wealthiest, and best educated populaces in the country, the local EPB boasts only three lawyers on its staff, and they are responsible not only for organizing enforcement but also for developing policy and bylaws.[88] (In contrast, New York State, with only a few million more people, boasts ninety-eight lawyers in the State Department of Environmental Conservation and an additional thirty-seven environmental lawyers in the attorney general's office.)

Nonetheless, in some respects, the court system has emerged as the strongest lever that SEPA and its local bureaus possess for ensuring that environmental laws are effectively enforced, and there have been significant success stories. In 2002, a court in the Xuanwu district in Beijing awarded a woman almost $15,000 for medical and psychological damages that she and her family suffered when her home was found to have levels of formaldehyde nearly nine times higher than permitted levels as a result of work done by a decorating firm to her apartment.[89] In a second case, in January 2003, forty-seven farmers in Zhejiang Province were awarded almost $900,000 by a local court for a decade's worth of economic loss due to water pollution and poisoned fisheries.[90] In a case that not surprisingly received substantial publicity in China, more than 400 elementary school students from Jiande, Zhejiang Province, filed a class action lawsuit against a plastics plant after students became ill from a toxic chemical leak from the plant. The students received a partial victory for their efforts: The plant was forced to pay their hospital bills, but it has yet to enact measures to prevent such an accident from occurring again.[91]

The courts have been especially important in ensuring that when firms refuse to pay assessed discharge fees for exceeding pollution limits or fines for other violations, the plaintiffs receive their compensation. In Xuzhou in 1994, for example, the Xuzhou No. 2 Iron and Steel Mill began operating a 350,000-ton blast furnace without any environmental impact assessment and without building facilities to dispose of its wastewater, slag, and airborne emissions. Thus, wastewater containing substantial amounts of sulfides was discharged into a nearby deserted mining pit and polluted the underground water. When the local EPB fined the mill, however, the mill's directors refused to pay. The EPB then took the case to court. The People's Intermediate Court of Xuzhou fined the mill 20,000 yuan (approximately $2,500) for excessive pollution discharge.[92] In the mid-1990s, the Huangpu district court in Shanghai also forced twenty-five polluting firms to pay their fees after they rebuffed the local EPB.[93]

In 1999, when I visited Shanghai, a local EPB official described to me a case in which his office had sent a bill to a company that owned a huge landfill site that was leaching dangerous chemicals into the ground. The company responded that it did not have the money to pay the fees. The Shanghai Environment Protection Bureau then sued in court and won.[94]

The courts have even begun to sentence enterprise officials to jail on environmental issues: In 1998, for the first time, the head of a polluting enterprise was sentenced to jail for two years for environmental crimes.[95] Since that time, both those directly responsible for the pollution and environmental protection officials responsible for the oversight of the polluting firms have been found negligent. In August 2001, for example, SEPA announced penalties against local government officials in eight provinces "for failing to . . . enforce national environmental protection laws."[96] Just one year later, the director and vice director of a county-level EPB in Shanxi Province were sentenced to prison for six and eight months, respectively, for failing to stop a chemical plant from discharging toxic waste into the drinking water system of a local village. The villagers had sent a letter of complaint to local authorities, and forty-nine of them had been hospitalized with serious illnesses from the water.[97]

Perhaps the most potent weapon in China's environmental protection enforcement effort is the advent of the nongovernmental environmental advocate. In 1998, Wang Canfa, a charismatic and hard-driving professor of law at the China University of Politics and Law, established the Center for Legal Assistance to Pollution Victims. The center trains lawyers to engage in enforcing environmental laws, educates judges on environmental issues, provides free legal advice to pollution victims through a telephone hotline, and litigates cases involving environmental law. The center's resources, however, are stretched thin: It must raise its own funds through private sources, undertake the investigation of the cases, and prosecute them.[98]

Top officials from SEPA commend Wang's work, but they note the limited capacity of his organization because it depends on contributions for financial support. Perhaps recognizing the limitations of SEPA, one official expressed the hope that Chinese citizens would not only support Wang's activities but also find their own ways of using China's environmental laws to protect themselves.[99] During Wang's trip to the United States in 2001, I had the chance to sit down to chat with him about his work, and he was quick to point out that not all local EPBs or courts are as supportive of his work as SEPA. Since plaintiffs typically approach Wang only after having exhausted the appropriate bureaucratic channels such as the local EPBs, environmental protection officials are often afraid that they will be held liable for inaction. In one case, written up in both the Chinese and U.S. press, Wang won $240,000 in damages from a liquor factory for eight fish farmers whose geese and fish were killed by polluted wastewater from the factory. The local government in this case had refused to help, afraid that it would be held liable for not taking action earlier.[100] In a separate case, Wang tried to file a suit against a polluting coal mine outside Beijing on behalf of seventy-one villagers. The court demanded that Wang file seventy-one separate suits in order to collect fees for each one.[101]

One promising, albeit underutilized, avenue for enhancing enforcement is the occasional alliance between the courts and the banking system. China's state-owned banks have become engaged in environmental protection through both fund-raising for environmental projects[102] and denying credit to firms that do not maintain pollution standards. In the mid-1990s, the People's Bank of China adopted a policy of refusing to extend credit to firms that did not correctly dispose of their industrial waste or that failed to meet state standards for environmental protection.[103] More common has been coordination between the courts and the banks to ensure that if firms do not pay their fines, their assets may be frozen and fines deducted from their accounts.[104] However, during interviews with Chinese environment officials in eight cities during 2000, I uncovered only one case—in Dalian—in which an official reported using this mechanism.

A separate case, reported by the Chinese press, helps illustrate why such efforts are rare. In this case, forestry officials in Shanxi Province sought a fine totaling $530,000 from a highway construction company for destroying forests along a 42-kilometer highway it built from Shanxi to Henan. Part of the problem was that forestry officials were not included in an interagency group planning for the highway. When one of the forest bureaus involved assessed a fine on the company, the company ignored it. The bureau then obtained a ruling to freeze the assets of the company from a district court. Only a few days later, however, a senior government official persuaded the court to release the company's assets.[105]

The institutional weakness of SEPA, the problems of the legislative and judicial organs, the power of the financial and planning agencies, and the relatively low level of central investment in environmental protection all have far-reaching negative implications for the center's ability to implement its policy mandates. Moreover, the decision to devolve authority for environmental protection to the local level as well as the nature of decision making at the local level have produced a situation in which environmental protection is spotty throughout the country. Only a few of the wealthiest locales with the most proactive leaders and strong ties to the international community have made real progress toward redressing the environmental degradation of the past and working to respond to the challenges of the future. Still, these areas offer hope that as China continues to integrate into the international economy, wealth increases, and officials become more educated as to the benefits of environmental protection for the continued prosperity of their region, more environmental leadership will emerge from local officials.

Devolution of Authority

In 1989, the Environmental Protection Law established the "environmental responsibility system" (*huanjing baohu mubiao zerenzhi*). In theory, this placed the responsibility for environmental protection in the hands of local officials, who together with local EPBs would work together to improve environmental protection.[106] Investment decisions, experiments with new policy tools, engaging the public, and establishing linkages with international actors all fell within the portfolio of these leaders.

Without a strong central apparatus, however, this highly decentralized system of environmental protection has continued China's prerevolutionary reliance on the environmental proclivities of individual local officials. When the mayor is environmentally proactive, income levels are high, and the city is tightly integrated into the international community, environmental protection has evidenced substantial progress over the past decade. When these conditions are not met—and especially when the mayor or governor is not committed to environmental protection—the EPBs are unlikely to be effective actors.

A province or city's wealth largely determines the resources it can devote to environmental protection. Thus, Shanghai with GDP per capita of $4,500 and Xiamen with $5,000 devoted approximately 3 percent ($2 billion) of their local revenues to environmental protection in 2002. In 2003, Shanghai's mayor announced that he expected to increase the share of local revenues apportioned to environmental protection to 5 percent. By contrast, a province like Sichuan, with a per capita GDP of $576 in 2000, devotes only 1 percent of its local revenues to environmental projects.[107]

Not surprisingly, in their study of the efficacy of water pollution management policies in the early 1990s, Scott Rozelle, Xiaoying Ma, and Leonard Ortolano found that success was greatest when there was a wealthy local government and a high proportion of state-owned firms.[108] In contrast, EPB officials in many of the poorer cities, such as Chongqing or Kunming, complain that SOEs are the most difficult to control because they are central to the health of the local economy.[109] EPB officials in Chongqing, who are responsible for cleanup of the pollution in the Three Gorges basin, have stated that SOEs, which are responsible for 50 percent of the waste generated in the region, cannot be fined, brought to court, or closed down because of their importance to the economy.[110]

High income levels also generally translate into higher levels of environmental education and a greater number of citizen complaints about environmental issues. In wealthier provinces, the number of complaints is up to five to seven times higher than in less-developed regions.[111] Complaints are important because they often inform EPBs of situations of which they are unaware. In Dalian, for example, 40 percent of the almost 2,000 cases investigated by the EPB stemmed from citizen complaints.[112] Major cities throughout China have set up hotlines so that local people can report environmental concerns; in the wealthy coastal region of Zhejiang, the provincial-level EPB received more than 830 phone calls during the first half of June 2001.[113] Environmental officials in poorer regions such as Chongqing, Yunnan, and Sichuan, in contrast, complain that poor citizen education and environmental awareness is a significant factor contributing to their difficulties in protecting the environment.[114] After Chongqing established a hotline in June 2002, however, officials received more than 10,000 phone calls within the first six months of operation,[115] suggesting that the problem may not be as much citizen awareness as the lack of a mechanism for transmitting this awareness to officials.

Wealth is not the only or even the most important determinant of how well a city manages its environmental protection efforts. Guangzhou, whose per capita GDP topped $5,000 in 2002, devotes only 1.5 percent of its local revenues to environmental protection; thus far, it has achieved far less than Shanghai or other coastal cities in addressing its water and air pollution challenges.[116] Although, perhaps in a sign of change to come, in 2003, Guangzhou announced a number of projects to improve its water quality.

Much of the difference stems from the commitment level of the mayors. Zhongshan, Shanghai, and Dalian—all cities recognized for their environmental protection efforts—have had mayors who have made the environment a top priority. These cities boast not only physically beautiful environs but also a positive trend line in their pollution statistics and cleanup efforts. Their EPBs also possess enough clout (thanks to mayoral support) to persuade other local bureaus to engage in cooperative activities. For example, in Dalian, the EPB has worked well with the local public security bureau to enforce the monitoring of auto emissions. Confident of mayoral support, some EPBs become more proactive in developing policy initiatives. For example, in 1997, the Shanghai EPB pushed to introduce unleaded gasoline to improve local air quality. It did survey research, assessed the investment needed, and reviewed the storage and transportation capacity of the city. After reviewing the evidence, Mayor Xu "simply decided" that Shanghai would introduce unleaded gasoline.[117]

Shanghai is also one of the few cities that has taken proactive steps to address its mounting water problems through price reform. The mayor enforced a price increase in tap water between 25 and 40 percent to fund water quality improvement programs and to make sewage self-financing.[118] Dalian took such steps only when confronted with dramatically reduced water supply as a result of the devastating drought in summer 2001. At that time, Dalian officials decreed that any family exceeding the official water limit per month had to pay ten times the usual price for water.[119]

Even when officials recognize the benefits of a more environmentally proactive course of action, concerns for social stability often prevent them from adopting such measures. Despite the fact that agriculture in China rarely employs the most efficient irrigation technologies, for example, water prices are often artificially depressed to encourage farmers to remain on the land. In 2000, when the government of the southern city of Foshan, outside Guangzhou, attempted to raise its water prices, the effort was defeated by the local people's congress, which feared social unrest from increased water prices. Even Beijing has failed to increase the price of its water significantly despite its growing scarcity. In November 1999, there were two price hikes; however, water still only costs 1.3 yuan (fifteen cents) per cubic meter for consumers and 1.6 yuan per cubic meter for industry. The real costs, however, should be closer to 5 yuan per cubic meter if Beijing expects to attract private investment to meet the demand forecast for supply and treatment facilities.[120] Only now that the prospect of the nearly $58 billion South-North Water Transfer Project is becoming a reality does the Beijing leadership appear to be taking seriously the necessity of tough new efficiency and pricing measures. In early 2003, Beijing announced a significant hike in the price of water to almost 3 yuan per cubic meter.[121]

China's top environmental cities also have extensive ties with the international community, including long-term, comprehensive cooperation efforts with the World Bank, Asian Development Bank, and countries such as Japan, Germany and Singapore. For example, Japan has an exchange program with Dalian's EPB, and the Japanese Overseas Cooperation Agency assists in urban planning, establishing environmental priorities, and pollution monitoring. Japanese companies, a major source of foreign investment in Dalian, are also considered environmentally friendly.[122]

Still, these wealthiest, most environmentally proactive cities present their own set of challenges. They often pursue environmental cleanup by exporting their polluting industries to points just outside the city limits. While this can be an effective strategy if the outlying regions are sparsely populated, population pressures in the major cities often drive the development of satellite communities in these outlying areas. In the past seven years, Shanghai has moved more than 700 factories outside the city limits. Guangzhou, too, is planning to move all of its cement factories outside the city during 2000–2005.[123] Zhongshan, another model environmental city located a few hours from Guangzhou, is surrounded by strip mining enterprises, petrochemical plants, semiconductor factories, and other polluting industries. Many of them are joint ventures with Taiwan.[124] During a drive with the vice mayor of Zhongshan in which he proudly recited the economic and environmental statistics of his beautiful city, I gently pointed out that the environmental achievements of the city might be even more impressive had they not been achieved on the back of outlying towns. While the irony of my comment seemed lost on the vice mayor, the local EPB director laughed heartily at my remark.

The Campaign Mentality

While China's leaders have devolved both authority and responsibility to local officials for environmental protection, they continue to intervene in environmental management through broad campaigns to address problems of nationwide importance and scope. They use campaigns to pressure local officials on the full range of macro-environmental threats: deforestation, desertification, and water scarcity. Campaigns, however, suffer from three major shortcomings: (1) they tend to be highly politically charged with significant investment up front but little follow-through past the stated target of completion; (2) they rarely consult local officials and businesses to engage them in the campaign; and (3) they often do not employ the best set of technologies or incentives to change behavior. Overall, these campaigns place extensive burdens on localities with little follow-through from the center and thus, often, fail to achieve the desired results.

In 1998, for example, after devastating floods of the Yangtze River, Premier Zhu Rongji announced a ban on logging for huge swaths of western Sichuan Province. Since that time, the ban has been extended to seventeen provinces, autonomous regions, and municipalities. Reports regarding the progress of the ban to date, however, are conflicting. Official Chinese statistics attest to success, with China's forest coverage increasing from 13.4 percent to over 16 percent during 1996–2001.[125] According to one report, "Logging has been reduced in Manchuria, Inner Mongolia, the northwestern Xinjiang territory and elsewhere. In all, 740,000 wood workers have been laid off. . . . China's timber production plummeted 97 percent from 1997 to 2000, when only one million cubic meters were produced."[126] According to this same report, Chinese loggers have offset the losses from the ban by clear-cutting wide swaths of forest in Burma,[127] and other reports suggest that China is becoming a significant importer of timber, legal and illegal:

> China has become, virtually overnight, the second largest importer of logs in the world, trailing only the United States. . . . (The volume of uncut logs arriving in China has more than tripled since 1998 to over 15 million m³ [cubic meters]). . . . About 40 percent of China's timber imports come from illegal clear-cutting. . . . Timber imports from Russia have risen from less than 1 million m³ in 1997 to nearly 9 million m³ in 2001, and [twenty percent] of this trade is believed to be illegal. . . . [China is] quickly becoming the leading destination for Indonesia's illegal logs.[128]

In addition, among the largest foreign logging companies in the Amazon basin, Chinese logging companies, along with those from Malaysia, were fined in 1998 for failing to comply with forestry laws; after failing to pay their fines, their projects were halted.[129]

But other reports suggest that imports, whether legal or illegal, have not had a salutary effect on domestic production. According to one researcher with the State Forestry Administration, the flood of imports has not helped to protect Chinese forests but instead has forced the price of timber down, encouraging domestic loggers to fell more trees to compete.[130] Indeed, with entire local economies dependent on the timber industry and well-known collusion between local authorities and business to undermine central regulations, it would be surprising for such a ban to elicit such a dramatic change in such a short time. Thus, there have also been reports of flagrant disregard for logging restrictions. An investigation conducted by the State Forestry Administration in late 1999 found that officials in many localities were ignoring the ban and issuing permits in excess of their authority. As a result, logging in many areas exceeded legal limits by more than 150 percent.[131] The UN Food and

Agriculture Organization statistics also indicate virtually no decline in Chinese timber production during the period 1997–1999. Chinese environmentalists, apparently not convinced of the success of the logging ban, have voiced their concern that at the current rate of timber production, China's forests will be depleted in about ten years.[132] Moreover, without supporting social policies, such as a social security system for the approximately one million lumberjacks who will be left jobless,[133] many are likely to return to their previous jobs. It may well be that as the economy continues to grow, and construction continues to skyrocket throughout the country, the increased imports are merely meeting the growing demand.

Yet there is little doubt that the center has attempted to assert greater control over the illegal logging. Toward the end of 2000, the State Forestry Administration announced a ten-year campaign to crack down on illegal lumbering; funding for the campaign was set at $11.6 billion. In 2001, the State Forestry Bureau reported that it already had pursued 2.2 million cases involving the destruction of forests over the past five years.[134]

Afforestation campaigns, too, despite their long history in China, have been plagued by inappropriate technology, poor oversight, and a failure to provide appropriate incentives for actors to change their behavior. In discussing the massive afforestation projects in Shandong, the Yangtze River Valley, and across China's northern provinces during the late 1980s, one survey by the Northwest Institute of Forestry in Xian revealed that "half of the reported national afforestation claim was false," and "the survival rate of planted trees was no higher than 40%."[135] More recently, in 2001, *China Environment News* reported that a Beijing-directed plan to plant 230 hectares (560 acres) of trees outside the capital in order to protect the capital's water supply had resulted in only about 70 trees being planted, and of those many had died.[136] Planting trees too close together, poor forest management, and planting trees without consideration for the viability of the tree species in the local ecosystem have all limited the efficacy of tree-planting campaigns.

In 2000, the government called on the people to counter the challenge of desertification, announcing a $725 million campaign to prevent and control desertification by adding new grassland and forest and increasing the vegetated area throughout the northwest. In some cases, resettled farmers and herders were directed to change the source of their livelihood from agriculture to tree planting.

Despite the promised assistance, however, some regional officials have already complained that Beijing has not come forth with the funds for these mandated reforestation efforts. In Inner Mongolia, the deputy Party secretary has claimed that Beijing has supplied neither the people nor the financial assistance necessary to reforest his region, which is already 60 percent desert. He has called on local business people to "invest in saving their homeland."[137] Even close to Beijing, where the desert is now only 110 kilometers away, villagers are unclear how to respond to the government's call to "grow trees, stop farming hilly land and confine livestock in pens to prevent ecological disaster." One villager, Zhou Qingrong, who already subsists on little more than $50 per year, had this response: "How can we make a living if we do that? Already we don't even have enough money to buy oil or salt."[138]

Even some Chinese officials are worried about the government's approach. One official in the Ministry of Agriculture's ecology section is concerned that "everything is going fast and there is no masterplan." The campaign approach "misses the step-by-step approaches needed in a well-rounded environmental package [including] planting grasses first to stabilize and enrich soil, then trees."[139]

Similarly, in 1997, the government initiated the "Three Rivers and Three Lakes" campaign to clean up the Liao, Huai, and Hai rivers and Tai, Chao, and Dianchi lakes. Government reports in 1999 touted the measures taken by the Kunming government to clean up China's sixth largest freshwater lake, Dianchi Lake:

> The Kunming city government banned the sales and use of detergent which contains phosphorus which was then discharged into the Dianchi Lake valley. It also closed down more than 20 enterprises found to be responsible for pollution. A project to dredge the bottom of the lake is also going on smoothly and should be completed by April this year.[140]

EPB officials in Kunming, however, report a very different story. In part, as they note, the sheer magnitude of the problem is overwhelming:

> Over the years, lots of industrial waste water has been emptied into the upper section of Dianchi. As a result, the lake's purification ability is very weak. No matter how much dredging we do, there will still be pollutants in the water. Because of years of abuse, the ecosystem is very fragile. Blue algae blooms are common.[141]

In fact, the sediment at the bottom of the lake contains a toxic mix of cadmium, arsenic, and lead. Even after absorbing $2 billion in loans and grants, most of the lake's water fails to achieve even grade V standard.[142] According to the deputy director of the Yunnan Provincial Institute of Environmental Science, Xu Haiping, it would take 20–30 years for the water in the lake to be as clear as it was in the 1950s.[143] Even that may prove to be optimistic as officials decry their ability to prevent polluters from continuing to pollute:

> We've banned phosphorus detergent and we're also promoting the use of natural fertilizer. This program hasn't really gotten off the ground. This is an area where we would like NGO help. Farmers have been opposed to the natural fertilizer, so there hasn't really been any progress. They need to be educated. . . . I'd estimate that approximately 70% of all waste discharge comes from chemical refineries . . . the most problematic polluters are state owned companies. We cannot take them to court.[144]

By fall 2001, local officials despaired of ever cleaning up the lake. According to Lin Kuang, of the Dianchi Regulatory Commission, "We just can't make the water clear."[145]

Perhaps the central government's most audacious bid to address both water scarcity and water pollution in China's northern provinces, however, is the South-North Water Transfer Project. This project, which was initially proposed by Mao Zedong in the 1950s, began construction in December 2002 and encompasses three canals, diverting up to 44.8 billion cubic meters of water from the Yangtze River to the Yellow, Huai, and Hai Rivers at an estimated cost of almost $60 billion.[146] While construction of at least two of the canals (the eastern and central) is not in doubt, the third (the western) requires such an enormous investment of technical and financial resources, that many believe it will never be completed.[147]

This massive project first had to overcome the serious resistance of a number of provincial leaders. During the discussion phase of the project, many officials in the southern provinces complained with good cause that Beijing was wasting a tremendous amount of water from easily addressed technical problems such as leaky toilets, and thus resented Beijing's looking south to meet its water needs. Even former Chinese president Jiang Zemin complained, "If a country can send satellites and missiles into space, it should be able to dry up its toilets."[148] At the same time, Hubei and Henan Provinces will be adversely affected: The project will require the resettlement of upward of 300,000 people from these provinces.[149] As discussed in chapter 7, China's experience with resettlement during its Three Gorges Dam project does not augur well for this process.

An even greater challenge for Beijing will be to ensure the cooperation of all the affected provinces in cleaning up the pollution that plagues much of the area along the proposed canals. By 2008, for example, government officials have called for the construction of 295 water pollution control projects along the eastern canal alone, along with the closure of many highly polluting factories, such as paper and pulp mills.[150] Other such large-scale campaigns to clean up rivers and lakes, as discussed above, however, have proved largely ineffectual.

The environmental impact of the canals themselves remains in question. There are widely varying analyses, including some that predict a largely positive impact for the eastern route,[151] and others that fear significant disruption of the local ecology, salinization of the soil that would cause the water quality in Shanghai to deteriorate, and damage to freshwater fisheries.[152] A Chinese official offered the most pessimistic view of the water transfer project: that the river diversion would not be able to transport sufficient water to solve the problem, especially during the winter months when the water was too low and there was not superfluous water flow.[153]

Even as Beijing has mobilized provincial support for the construction of the South-North Water Transfer Project, it will require a commitment of substantial financial and political resources to ensure its success. Resettlement, pollution control, and limiting the ecological impact all will require continual oversight by Beijing. The campaign approach to environmental protection, while useful for garnering official attention and mobilizing public support for a particular environmental challenge, has little chance to succeed. It emphasizes above all a crisis mentality and grand, sweeping gestures. Instead, China's leaders will have to undertake the careful planning, long-term investment, and closely monitored implementation that typically eludes them as they seek to ensure that this grand-scale development project and cleanup campaign fulfills its goals.

The Failures of Reform

Deng's early steps toward institutionalizing a bureaucracy and legal framework for environmental protection during the 1970s and 1980s, as well as later initiatives under his successors Jiang Zemin during the 1990s and into the twenty-first century under Hu Jintao, have not been able to meet the rapidly multiplying set of environmental challenges that have emerged in the wake of the economic reforms. In virtually every respect, China's environment overall has deteriorated, contributing to additional social and economic problems. In particular, China's central environmental protection bureaucracy remains weak, forced to negotiate with more development-oriented ministries and commissions from a position of lower status within the central bureaucracy and unable to enforce its directives through a strong chain of command to local EPBs.

At the same time, China's formal environmental protection institutions have all taken important steps to enhance their efficacy. They have improved the process by which they develop new policies and draft new laws, experimented with new policy approaches, and developed a coterie of talented and highly qualified experts throughout the policy, legal, and judicial arenas.

Too, while the devolution of authority to local officials has left many regions suffering from lack of leadership or lack of resources or both, it also has produced some striking examples of success. Local initiative in Dalian, Shanghai, and Zhongshan, for example, has earned all three cities recognition for their environmental achievements.

Moreover, as chapters 5 and 6 explore, greater change may be forthcoming from forces outside the government. China's reforms also have introduced nongovernmental actors, including international and domestic NGOs, the media, and multinationals as important sources of environmental change, providing access to new policy approaches, new technologies, and funding opportunities. Often at the margins, but sometimes in critical ways, these actors are helping to push China's environmental protection capacity beyond the limits of its formal institutions.

CHAPTER FIVE

THE NEW POLITICS OF THE ENVIRONMENT

Citizens should have the right to shout out that the emperor has no clothes
on whether they are right or not. Without democracy there can be no way
to approach truth. History without clear conclusions will only repeat its
blind and restless past. . . . I found the chief guarantee of nature protection
to be the practice of democracy. Without real democracy there can be no
everlasting green hills and clear waters. I am convinced that nature conservation is a cause for a whole nation.

Tang Xiyang, *A Green World Tour*

China's leaders have come to understand that failure to protect
the environment incurs significant social and economic costs,
and they are eager to find a means to reconcile their desire to
achieve both unimpeded economic growth and improved environmental protection. As we will see in chapter 7, the path they have
elected to follow is the one taken by the East Europeans and Asians
over a decade ago, establishing an environmental protection bureaucracy and legal system and supporting the existence of environmental nongovernmental organizations (NGOs). Beginning with the
launch of the first environmental NGO, Friends of Nature, in 1994,
China's leaders have opened the political space for popular participation in environmental protection, permitted the establishment of
NGOs,[1] encouraged media investigations, and supported grassroots
efforts.

At the same time, China's leaders have resisted widespread imple-

mentation of tough economic policies, such as raising prices for natural resources or closing polluting factories, that might engender social unrest. By promoting the growth of environmental NGOs and
media coverage of environmental issues, the Chinese leadership
hopes to fill the gap between its desire to improve the country's environment and its capacity to do so.

This trend in inviting greater political participation in environmental affairs represents a much more widespread phenomenon in
state-societal relations: the emergence of nongovernmental associations and organizations to fill roles previously occupied by national
or local governments and state-owned enterprises. The rapidly expanding market economy has diminished the capacity of the state to
regulate the economic and social activities of the Chinese people.[2]
At the same time, the state has intentionally abandoned its commitment to the "iron rice bowl," in which the state provided for all
the basic needs of society. The Chinese leadership no longer desires
responsibility for much of the burden of social welfare that it has
borne for the past fifty years, including the provision of housing,
medical care, education, pensions, and environmental protection.
Instead, it is encouraging the private sector to respond to the social
welfare needs of the public, opening the door to the emergence of
civil society, permitting greater openness in the Chinese media, and
encouraging the Chinese people to assume greater responsibility in
managing their own economic and social affairs.

This is an enormous gamble. As the Soviet and Eastern Bloc experience shows, social forces, once unleashed, may be very difficult to
contain. While the Chinese leaders seek the antidote to a dysfunctional political and economic system, they are also desperate to
avoid the political turmoil that convulsed the Eastern Bloc countries and ultimately transformed those political systems. Chinese
officials have expressed concern that social groups could contribute
to "peaceful evolution" or call for the transition of China to a west-
ern-style democracy.

China's leaders therefore have been careful to circumscribe both
the number of NGOs and the scope of their activities. Yet there is
evidence—nascent still—that the Chinese leadership will prove no

more adept than the East Europeans at managing reform while avoiding revolution. As was the case in some of the countries of Eastern Europe and republics of the Soviet Union, as well as countries in Asia, environmental NGOs in China are at the vanguard of nongovernmental activity. Thus the question is not only whether nongovernmental actors can shape the future of environmental protection in China but also whether they may play a role in effecting broader political change in the context of the ongoing transformation of state-societal relations.

Tightrope Act: NGOs and the Response from Beijing

The Chinese government's decision to alter the social contract that has governed state-societal relations for the past several decades has been met with both enthusiasm and concern by the Chinese populace. While people welcome the opportunity to exercise greater freedom of movement, association, and so on, they are also wary of the new responsibilities entailed. Public opinion polls, for example, reflect a growing concern with social issues emerging from the reform process.

The environment is one of the most important of these issues. In a 2000 poll, environmental degradation and protection was the number one concern cited by three thousand urban Chinese in ten cities, followed by unemployment, children's education, social stability and crime, corruption, economic growth, and social security. (During the previous five years, the top concerns were social stability and crime, along with unemployment.)[3] In a 2002 Internet poll of thousands of Chinese citizens conducted by the *People's Daily*, the environment ranked tenth as a concern, behind other social concerns such as official corruption, the impact of World Trade Organization (WTO) accession, and peasants' low incomes.[4]

In response to these growing social welfare concerns and demands, a wide range of social organizations—NGOs as well as government-organized and supported efforts—has emerged, addressing issues as wide-ranging as domestic violence, job retraining, and environmental protection. According to the Ministry of Civil Affairs, which oversees such organizations, by 2002 there were 230,000 officially registered NGOs.[5] A broader definition of social organizations that includes all types of citizen-based organizations and economic associations yields a number over one million. And if one attempts to account for NGOs that are not registered with the government, the number by some accounts is as high as two million.[6]

The Chinese leadership has generally welcomed the efforts of these organizations. They are somewhat reminiscent of organizations that appeared during the republican period, when religious groups, literary societies, and relief organizations emerged and prospered, at least until the demands of mounting a war against Japan in the 1930s led to a government crackdown on independent societal organizations.[7] During the 1980s, in the wake of Deng Xiaoping's initiation of economic reform, some of the same kinds of relief organizations emerged, as did a few groups calling themselves NGOs. Some of the latter even became active in policy research and advocacy.[8] Again, a crackdown followed, this time on the heels of widespread protests for democracy and political change.

China's leaders today remain worried about the potential for the current generation of NGOs to serve as a locus for intellectual, worker, peasant, or general societal discontent. They already have to contend with substantial organized urban and rural unrest, dissident labor unions, a resurgence of triad (i.e., China's traditional criminal gangs) activity, and growing participation in religious organizations that are not sponsored by the state. The controversial Qigong movement Falun Dafa,[9] which embraces Chinese from all sectors of society and has demonstrated the capacity to mobilize thousands of people without being detected by the Chinese public security apparatus, epitomizes the type of challenge the government fears.

Such concerns have produced a duality in the government's approach to the NGO sector. On the one hand, NGOs fill societal needs and earn international praise. In 1996, at the Fourth National Conference on Environmental Protection, Song Jian, then head of the State Environmental Protection Commission (SEPC), stated, "As for various environmental protection mass organizations which are concerned with environmental protection undertakings, we should actively support them, strengthen leadership, and guide their healthy development."[10] In 1997, in his speech at the fifteenth Chinese Communist Party (CCP) Congress, then CCP general secretary Jiang Zemin echoed Song's sentiment, highlighting the necessity of "cultivating and developing social intermediary organizations."[11] During former U.S. president Bill Clinton's visit to China in June 1998, the Chinese government sponsored a roundtable for President Clinton with environmental NGO leaders.[12]

On the other hand, during 1995–1997, the government placed a two-year moratorium on the registration of new NGOs, and in September 1998, the State Council issued a set of restrictive *Regulations for Registration and Management of Social Organizations*.[13] Previous NGO guidelines had been vague.[14] These new regulations, however, required all NGOs to reregister, stating that, in the future, all such organizations would have to receive approval from a government organization before applying for NGO status and that there would be no appeal process if an organization's application were rejected.[15] They also stated that a national-level NGO would have to "prove [that it had] a 'legitimate' source of funding, raise at least 100,000 yuan [$12,000] and comprise of [sic] more than 50 individual members."[16] Local NGOs would have to raise 30,000 yuan ($3,750).[17] Moreover, anyone who had ever been deprived of political rights, such as a former political prisoner, would be barred from participating in an NGO.[18]

Why the mixed message? One Chinese scholar assessed the government's concern over the NGO sector this way: "The Party knows from its own experience that it is possible to start a mass movement capable of overthrowing a government from just a small group of about a dozen people. . . . As such, control is vital. They feel they cannot allow any non-official group to gain ground in society for fear it will grow into a potentially threatening movement."[19] The emergence of a particular NGO, the China Democratic Party (CDP), highlighted just such potential. Using the Internet to communicate among a small membership scattered throughout the country, the CDP tried to register as an NGO in one-third of China's provinces and autonomous regions. This was despite its openly stated mission of establishing itself as a political alternative to the Communist Party, which was clearly a subversive activity outside the boundaries of acceptable NGO behavior. Beijing launched a crackdown. Within several months of its inception, the CDP was banned, and there was an aggressive manhunt to arrest those linked to the organization.

Thus, while permitting the NGO sector to develop in China, the government has been vigilant about limiting the range of its activities. In addition to the formal regulations, it retains a number of mechanisms by which it can effectively shut down an NGO. A Public Security Bureau circular issued in 1997 sets out three avenues by which local officials can effectively neutralize a "troublesome" NGO: (1) the sponsoring organization can cease its support; (2) the NGO can be closed down for financial reasons; and (3) key leaders of the NGO can be transferred to other jobs that leave them little to no time for outside work with the NGO. Anthony Saich, former Ford Foundation representative in Beijing, has noted that all three tactics have been employed. The Women's Hotline in Beijing (which provides counseling and advice to women), various NGOs in Shanghai, and "Rural Women Knowing All" (a group of activists concerned with women's issues in rural areas) have all been subverted through these methods.[20] According to Saich, who worked extensively with Chinese NGOs, the government is determined to maintain a high degree of Communist Party and state control over NGO activities.[21]

A revised set of draft regulations for NGOs was supposedly due to be released by the end of 2002 or immediately after the March 2003 National People's Congress (NPC); however, it continues to be delayed. One purported reason for the delay was the emergence of the Falun Dafa movement and likely ongoing debate within the Communist Party over how best to continue support while preventing another Falun Gong–like group from appearing.[22]

The government has also pursued development of its own NGOs, referred to as government-organized nongovernmental organizations (GONGOs). SEPA, for example, has three GONGOs under its auspices: the China Environment Science Association, the China Environment Protection Industry Association, and the China Environment Fund. In some cases, GONGOs serve as a resting place for former government officials and staff when their agency is downsized during a period of bureaucratic reform such as 1998 and 2003. They may also serve essentially as front organizations for government agencies. In fact, there is a state-directed effort to train government bureaucrats in the establishment of NGOs.[23] Some of these efforts are a legitimate means of establishing a cooperative venture with a foreign counterpart. For example, the Chinese government set up an NGO to work with the U.S. NGO, the National Committee on U.S.–China Relations, to work on land use planning for the Ussuri watershed on the border between China and Russia.[24]

Many other such GONGOs, however, are simply tools by which government agencies may take advantage of the desire of foreign governments and international NGOs to support the NGO movement and civil society in China. One example is the Chinese Society for Sustainable Development, which is linked to the Chinese government's Agenda 21 office and designed to attract money from western counterparts interested in working with NGOs to develop and implement projects directed toward sustainable development. Despite one official's claim that the NGO is a "genuine" NGO, the Society is headed by the director of the Agenda 21 office, staffed by Agenda 21 officials, and located on the fifth floor of the building in which the Agenda 21 office is housed.[25]

Over time, some GONGOs are likely to evolve into genuine NGOs, having already become significantly independent in their fund-raising or membership; there are also instances in which GONGOs and NGOs develop strong and mutually beneficial relationships, or in which GONGOs act as a bridge between government and genuine NGOs. As Wu Fengshi suggests in her excellent analysis of the community of environmental GONGOs, this evolution in the nature of GONGOs may have a profound impact in strengthening the development of green civil society in China, moving many of them well outside the boundaries initially established by the government.[26]

Notwithstanding the government's efforts to manage the activities of GONGOs as well as NGOs, the environmental NGO community in China has become adept at expanding the boundaries of the permissible while avoiding government censure. While NGOs do not explicitly violate government policy, they have achieved a level of national integration and potential for coordinated action against which the government has explicitly attempted to guard. For example, although NGOs are technically forbidden to establish branch offices, many province-based NGO leaders have been trained by or been members of Beijing-based environmental NGOs before setting off to establish their own organizations. There is now a growing network of provincial and local NGOs throughout China whose leaders have all received training or guidance from a dozen or so Beijing-based NGO activists. Moreover, the advent of the Internet has meant that environmental student groups as well as professional NGOs communicate with great ease, sharing both technical data and administrative knowledge on how to raise funds and increase membership. Also, some NGOs with national-level status have broad mandates such as conservation and protection of biodiversity and undertake activities throughout the country, although they are technically based in Beijing. They thereby have avoided the geographic stricture altogether and have members spread throughout the country.

Such circumventing of the spirit of official policy has not prompted a crackdown, primarily because these environmental NGOs have yet to challenge central government policy. Thus far, the government and environmental NGOs have operated under the tacit agreement that their mission and work are mutually reinforcing. Yet a closer examination of the environmental movement reveals that it has evolved into three distinct, albeit closely linked activist efforts.[27] Each of these strains within the movement has its own set of goals and means of operation, not all of which coincide with those of the government. Thus, the government–NGO alliance may be a tenuous one.

1. *Conservation.* The first and largest group of environmental activists devotes its time to nature conservation and species protection. These leaders, such as Liang Congjie from Friends of Nature, Yang Xin from Green River, Wang Yongchen from Green Earth Volunteers, and Xi Zhinong from Wild China, form a loose network of

committed environmentalists throughout the country who often work closely together.

While Beijing is supportive of the efforts of these NGOs, local officials often resist their involvement to protect entrenched local political and economic interests. As time has passed, some NGO leaders within this group are also beginning to raise questions concerning central directives. For example, some NGO leaders have pressed the Chinese leadership to include SEPA in the group of government ministries tasked with encouraging development in China's interior provinces in an environmentally sustainable way. While Beijing has been generally receptive, such lobbying is still at a nascent stage and has been done quietly at only the highest levels.

2. *Urban Renewal.* A second approach to environmental activism focuses on issues related to urban renewal. Liao Xiaoyi, the founder of Global Village of Beijing, Geng Haiying, a doctor in Dalian, and others who are concerned with urban issues are disinclined to challenge the center and more likely to try to work in a consensual fashion with local governments. Indeed, the nature of their work—recycling programs, energy efficiency, and environmental education—often demands at least passive assistance from local officials.

3. *Democracy.* Finally, the third group of environmental activists has interests and goals that exist well outside the boundaries for NGO activity established by the central government. These few but powerful voices include such renowned figures as Dai Qing, Tang Xiyang, and He Bochuan, who have clearly articulated the philosophical link between effective protection of the natural environment and the need for democracy. Some come to environmental protection as an outlet for their democratic leanings; others come to democracy from their belief in what it can do for environmental protection. While they themselves have not established their own NGOs, through their writings and activism they have exerted a profound impact on the next generation of Chinese environmental NGO leaders. Even those leaders of environmental NGOs in China that publicly eschew discussion of such issues privately acknowledge the necessity of greater openness for environmental protection.

Furthermore, underpinning all environmental activism in China are the Chinese media, which have been a critical partner to the NGOs in publicizing their accomplishments, working to expose corrupt local environmental practices, and informing and educating the Chinese public. Many of the brightest and most promising environmental activists of the next generation, such as Wen Bo, Hu Kanping, and Hu Jingcao, boast both environmental and journalistic credentials and may well revolutionize the manner in which Chinese NGOs pursue their goals in the years to come.

Three Pioneering Environmental Activists

The contemporary intellectual roots of environmental activism in China took hold during the mid-1980s and early 1990s with the writings of two prominent journalists, Tang Xiyang and Dai Qing, and one noted scholar, He Bochuan. While none of the three has established an NGO, each has had a profound impact on the evolution of the environmental movement.

Tang Xiyang

Revered by many in China as a spiritual and philosophical leader of the environmental movement, former *Beijing Daily* reporter Tang has had a tremendous impact on the ideas and thinking of environmental NGO leaders within China. Tang was born in 1930 in Hunan Province and graduated from the foreign language department of Beijing Normal University in 1952. Within five years, like thousands of other intellectuals who heeded Chairman Mao's call to let "a hundred flowers blossom and a hundred schools of thought contend," he came under political attack during the anti-rightist campaign, which condemned all those who had voiced their dissenting opinions openly. During the Cultural Revolution, Tang's wife was beaten to death by the radical, overzealous Red Guards, who were committed to rooting out all potential political opposition. Tang turned to nature as a refuge:

Driven into a corner, I could think of only two ways to relieve my suffering: kill myself or kill somebody else. Nature saved me. . . . Instead of twisted faces, hypocrisy, ruptured relations, and unending, unavoidable nightmares, nature offered the most sublime words, the most beautiful music, the purest emotions, perfect philosophy. . . . We spoke no words, yet we understood each other. . . . The more I love her, the more I understand her and her suffering, which is more terrible and intense than mine. . . . Now I no longer appeal for myself, but for her.[28]

After Deng Xiaoping's accession to power, Tang was rehabilitated in 1980, and he joined the Beijing Museum of Natural History as the editor of the museum's magazine, *Nature.* The journal became a vehicle for his and others' musings on issues such as species protection.

In 1982, Tang met an American cultural education specialist, Marcia Sparks, and the two began to travel together to national parks. They fell in love and were married. In 1996, Tang and Sparks established the first Green Camp, which served both to train future environmental activists and to draw attention to the plight of some of China's most precious resources and endangered species. Tragically, Tang realized their dream alone. Early on the morning they were scheduled to travel to Yunnan for the first Green Camp, Marcia succumbed to a long battle with cancer. She had placed herself in a hospital, strictly in opposition to her religious beliefs as a Christian Scientist, so that Tang would not feel compelled to stay with her and care for her at home but rather would forge ahead with the Green Camp. With as many as thirty college students, Tang journeyed to northwest Yunnan to study and work to protect snub-nosed monkeys. In 1997, Tang's Green Camp traveled to southeast Tibet to protect the primitive forests, and in 1998 they visited the Sanjiang Plain in Heilongjiang Province to protect the wetlands. Many of the students who participated in these Green Camps have returned to their homes in other cities, such as Shanghai and Nanning, to establish their own green camps.[29]

Tang's greatest influence on the thinking of Chinese youth, however, is probably through his 1993 book, *A Green World Tour.* The book—a personal account of Tang's eight-month-long travels through nature reserves and encounters with various people in countries in Europe, the former Soviet Union, Canada, and the United States, as well as in China—has inspired many Chinese to take steps to improve the environment, such as investing their own money to establish a nature preserve.[30] But Tang himself believes that much more must change for the environment to be protected in China. He comments,

> If my trips abroad can be compared to the journey to the West of the Tang Dynasty priest Sanzang, whose mission was to bring back Buddhist sutras, then what I am after is "the green sutra." I found the chief guarantee of nature protection to be the practice of democracy. Without real democracy there can be no everlasting green hills and clear waters. I am convinced that nature conservation is a cause for a whole nation. It won't do to depend on a wise emperor or president. Hundreds of millions of people must realize and show concern for this problem. When they all dare to speak and act, the emperor or president has to do something; otherwise he cannot continue in office. After visiting many countries and observing others' attitudes, I feel democracy is necessary to the protection of nature.[31]

The underlying message of the book, therefore, is a sharply political one. In response, the government censors have stepped in, deleting from both the 1993 Chinese and the 1999 English language editions Tang's ruminations on the Cultural Revolution, Tiananmen Square, and the imperative of democracy for China. He has countered the censors by amassing a large number of copies of the book at his home and reinserting the censored portions by hand. Many of his views run counter to the official line:

> The Tiananmen Square incident has left a very deep impression on me. All the [Chinese] newspapers, journals, radio and TV programs changed to uniform words and tone almost overnight, as if the happening involving a million people that had taken place the night before was unreal. A downright lie became 100 percent truth. In an

instant, intelligent men devoting themselves to the cause of democracy became criminals, while those engaged in repressing the masses were honored as "Defenders of the Republic." . . .

Let's compare conditions in Eastern Europe. When the peoples of Poland, Hungary and Eastern Germany awakened to the necessity for reform, reform became an irresistible tide. However, when the Chinese people awakened and began to act, they induced bloody suppression. That reminds me of various farmers' rebellions in former eras. Even if some of the rebellions were successful, the result was simply a shift in rulers, the fruits of success never passed into the hands of the common people. I'm also reminded of the burning of books and burying alive of scholars by the Emperor Qinshihuang, the state examination system started in the Tang Dynasty, and the various political movements started by Mao Zedong aimed at punishing intellectuals, which I experienced myself. In such a cultural environment, prolonged, closed, dictatorial, lacking in democracy, domestically and in interflow internationally, it was very difficult to develop science and technology as well as a tradition of loving nature.[32]

When asked directly, "What are China's biggest environmental problems?" Tang answers, "Democracy. If you don't have democracy, you can't have real environmental protection."[33]

He Bochuan

Tang's *A Green World Tour* remains in print in China, albeit in censored form. But distribution of one of the most controversial treatises, He Bochuan's *China on the Edge*, was halted a year after it was released.

Published in 1988, when the reform movement was in full swing and political openness in China had reached new heights, the book describes a grim environmental situation in China and predicts an even grimmer future unless radical political and economic reform are undertaken. The author, He Bochuan, was a lecturer in the department of philosophy at Sun Yat-sen University in Guangdong. In 1983, he contributed a section on education to the prestigious collection of essays *China in the Year 2000*. Using this piece as an intellectual launching pad, He wrote a number of articles on the related issues of population, ecology, energy, and economics that eventually became *China on the Edge*. By the time publication was halted in 1989, in the wake of Tiananmen, it had already sold more than 400,000 copies.[34] The book was, in fact, widely distributed among students during the Tiananmen Square protests.[35]

At the time, He's contribution was unique not only in its critique of environmental practices but also in its clarion call for political and economic reform. He linked issues such as inefficiencies in water supply to irrationalities in the economic system, noting, "These economic difficulties are a major factor in environmental problems because they create the inefficiency, waste, lack of innovation, and ineffective use of capital which compound environmental damage and obstruct solutions to it."[36] He later broadened his critique to include a more sweeping call for political reform: "until reform of the political system achieves concrete results, and particularly in political democratization, liberalization, and legalization, and systems for enforcing them, it will be impossible to devise an overall plan for China's current reforms."[37] Had the book continued to be sold, it might well have had the same impact as Rachel Carson's 1962 book *Silent Spring*, which is widely credited with catalyzing the environmental movement in the United States.

Today, as a Professor at Sun Yat-sen University, He continues to decry China's approach to development and environment, noting that "China is following an approach to development that measures progress only through economic growth, while ignoring the disastrous effects on the environment. . . . Economic growth is supposed to solve the problem of poverty, but it causes so much environmental destruction that poverty continues and development is undermined. It is a vicious circle."[38]

Dai Qing

Around the same time, in the mid-1980s, Dai Qing began the long personal odyssey that would transform her from an inquisitive reporter into a political pariah at home and an acclaimed human rights and environmental activist abroad.

In 1986, while a reporter at *Guangming Daily*, Dai attended a conference in Beijing on the Three Gorges Dam. A small group of Chinese scientists had organized the meeting to report on their environmental concerns about the dam. While Dai did nothing with the information at the time, she maintained an interest in the issue. During a visit to Hong Kong one year later, she encountered a writer named Lin Feng, who provided her with information on the dam from Hong Kong newspapers over the next few years.[39]

In early 1989, when the NPC and the Chinese People's Political Consultative Conference were scheduled to discuss the proposed dam, Dai attempted to inform the debate by arranging for several prominent journalists to interview the scientists she had met in 1986. When it became clear that no newspapers or journals would publish the interviews, she arranged to have the collection published by the Guizhou People's Publishing House,[40] with the same editor, Xu Yinong, who had published He Bochuan's book. Guizhou People's Publishing House, an obscure press in one of China's poorest provinces, likely operated under the central government's radar screen. Generally, officials permit far greater latitude in political expression and economic reform away from the capital of Beijing.

Dai's book, *Yangtze! Yangtze!*, was quickly acclaimed in the West and banned within China (although 25,000 copies remained in circulation) for its articulate and scientifically based evidence opposing construction of the Three Gorges Dam. The book includes essays by journalists, scientists, and well-known political figures within the CCP's leading group responsible for evaluating the dam's feasibility who dissented from the government's final published report, as well as by others involved in assessing the dam's viability.[41] Implicit and occasionally explicit throughout the book is a critique of the lack of political openness in the environmental assessment process of the dam, the silencing of dissident opinions, and the blatant disregard for truth displayed by the leaders who favored the dam.

One of the most telling criticisms comes from the highly respected former personal secretary of Mao Zedong, Li Rui:

[D]ue to the practice of relying on the leader's personal experience and will in decision making, we have yet to establish a complete scientific and democratic decision making process. . . . There are, however, some encouraging signs of improvement, such as when different views can be seen in the newspapers and many more scientists and specialists have the courage to voice their different opinions. But we are still in a transitional period between "rule of men" and "rule by law."[42]

Yangtze! Yangtze! may have represented the reform era's first public lobbying effort involving intellectuals and public figures in an attempt to influence a governmental decision-making process.[43] While the effort failed, its impact was also unprecedented in Chinese history: three years later, in 1992, when the NPC finally voted on the issue of construction of the dam, one-third of the delegates voted no.

Dai's opposition to the dam is as much political as environmental. As she once commented, "I got involved in this issue not for scientific reasons, but to promote the freedom of speech."[44] Her voice holds weight in China because of her family background. Her parents were killed by the Kuomintang (KMT) during China's civil war between the forces of Chiang Kai-shek's KMT and the Communists, and they were officially proclaimed "revolutionary martyrs." After their death, she was raised by Ye Jianying, one of Mao Zedong's chief lieutenants during the war, and then a long-standing member of the Politburo, who also served as a defense minister and then chairman of the standing committee of the NPC. Despite her family connections, Dai was forced to pay a steep price for her views: In 1989, she was sentenced to ten months in the maximum security Qincheng prison. The reason for her imprisonment was ostensibly her role in trying to broker a compromise between the students and the government during the Tiananmen crisis. However, she believes that the real reason for imprisonment was her vocal opposition to the dam.

Officially, Dai is not permitted to publish, to participate in any organizations, or to have a job. When she attempted to participate in a tree-planting project, the organizers were told that her name had to be removed from the list.[45] Still, she stays busy speaking throughout the world, and in 2003 completed a book about her experiences, *Prison Memoirs and Other Writings,* to be published in the United States.

Tang, He, and Dai represent the contemporary intellectual and political roots of the environmental movement in China. They, themselves, have not established environmental NGOs. Yet their ideals resonate with many of those who have. The current leadership of the environmental NGOs is at once emboldened and cautioned by the experience of these three writers and thinkers. And Dai and Tang especially continue to involve themselves heavily in environmental efforts. The environmental NGO community is small enough that most of the leaders are familiar to one another. As Tang Xiyang states, "All of the movement people know one another."[46]

The Conservationists

The largest, best funded, and best organized of the environmental NGOs in China are those that focus primarily on species and nature conservation and environmental education. Many of the leaders of these NGOs are intellectuals—historians, journalists, and university professors—who combine a reverence for nature with a sense of civic duty and political activism. Species protection also has been a politically acceptable issue to tackle because, at first blush, it does not challenge the deeply entrenched political priorities placed on urban and coastal modernization. Rather, species protection involves illiterate peasants, loggers, or officials in poorer, more remote regions of the country, where species still exist but are on the verge of extinction.

The conservation-oriented NGOs often work with national authorities to counter local officials' failure to implement central policies. As a result, these environmental NGOs have clashed with powerful local interests, sometimes with life-threatening consequences. More recently, they have even begun to challenge, in modest ways, the decisions of central authorities, suggesting that success may be emboldening them.

Foremost among these conservation-activists is Liang Congjie, who founded the first environmental NGO in China, Friends of Nature. Indeed, the birth of the contemporary environmental movement in China might be dated to the early 1990s, when a group of students and scholars searching for an activist outlet in the post-Tiananmen period approached Liang, an unassuming and highly articulate historian, with the idea of establishing an environmental NGO.

Liang's distinguished reform lineage made him a natural choice. His grandfather was the famous Qing dynasty reformer Liang Qichao, and his father, Liang Sicheng, was a renowned architect in Beijing, who attempted (unsuccessfully) to preserve much of the old city of Beijing during Mao's time. In addition, he holds a position in the Chinese People's Political Consultative Conference, a formal government body with a notable reform orientation, affording him a political platform as well as a degree of political protection that less well-positioned Chinese would not have.

While Liang doubted his ability to start such an organization, given his lack of expertise in environmental affairs and China's lack of tradition in nonstate organized activity, he consulted with some friends, such as Dai Qing, who encouraged him to give it a try. Liang was also inspired by the idea of a Chinese environmental NGO, having heard of an expatriate Chinese environmental group based in Boston, Green China. Perhaps most important, a vice director of the National Environmental Protection Agency (NEPA) had suggested to Liang that he use his voice to accomplish something positive in environmental protection.

With the support and guidance of his friends, Liang drafted a mission statement and charter for an environmental NGO. Rather than emulate Greenpeace, with its confrontational and often contentious tactics, he decided to emphasize environmental education. Mission statement in hand, Liang approached NEPA to serve as his NGO's "mother-in-law" (*guakao*). However, NEPA had already established its own GONGO, the China Environmental Cultural Association. Officials within NEPA suggested that Liang instead invite the Chinese Academy of Culture (*Zhongguo Wenhua Shuyuan*), of which Liang was a key member, to oversee the organization. With the Academy's support, Liang registered the NGO as the Green Culture

Sub-Academy on March 30, 1994.[47] By the end of 1994, Liang had received the government seal from the Ministry of Civil Affairs.

Even prior to receiving formal government approval for his organization, Liang held a first organizational meeting on June 5, 1993. About sixty people attended, primarily Liang's friends and friends of friends.[48] While Liang's friend Dai Qing herself was not permitted to participate because of her outspoken criticism of the government's plans to build the Three Gorges Dam, her friends attended the meeting.

The group focused its initial efforts on environmental education, publishing popular science-oriented books on environmental protection[49] and involving the public in conservation efforts such as tree planting. Soon, however, Liang expanded his group's activities well beyond the original mandate of environmental education to take on more controversial issues, including protection of species, such as the Tibetan antelope and Yunnan snub-nosed monkey, and preventing deforestation. The membership also grew rapidly, reaching 1,000 people by 2003.

While there are no branches of Friends of Nature in other cities—both because of the government's restrictions on such branches and because Liang is afraid of losing control of branch practices—several members of Friends of Nature have left Beijing and established their own environmental NGOs elsewhere. For example, renowned environmentalist and nature photographer Yang Xin, a former member of Friends of Nature, has established Green River Network in Chengdu, Sichuan Province.[50] Friends of Nature member Tian Dasheng also founded Green Volunteer League in Chongqing. Moreover, the NGO has spawned numerous offshoots that focus on specific aspects of environmental and wildlife protection, such as bird watching and animal rescue.[51]

Liang has become nationally and internationally renowned for his work, frequently traveling abroad and winning plaudits both domestically and internationally for his efforts. The Chinese government, for example, has awarded him the Giant Panda Award and sponsored him for a United Nations Environmental Programme Global Village Award. SEPA, especially, has appreciated Liang's work in calling to account local officials who ignore center directives. For example, in

1998, at a meeting sponsored by the Hainan environmental protection bureau (EPB), Liang challenged a local EPB official's account of Hainan's success in combating deforestation. A SEPA representative who had also been present later thanked Liang, stating that in his official capacity he could not so directly contradict the word of a local official, especially in a public meeting.[52]

Despite such acclaim, Liang maintains a modest demeanor and continues to work out of cramped offices in downtown Beijing. His ambition to improve China's environmental situation, however, remains large. During a visit with Liang and his wife in Beijing in 2000, he described to me the ways in which he and other NGO leaders had begun to try to reform Beijing's policies. For example, in late 1999, when the government announced its campaign to "Open the West," one of the five major tenets was "ecological construction." Yet SEPA was not among the twenty-odd ministries involved in advancing this policy, casting serious doubt on the government's real commitment to environmental protection. Liang, in his capacity as a member of the Chinese People's Political Consultative Conference, wrote a letter to the State Council calling on it to include SEPA in the leading group.

Liang also told me that he was planning to tackle new issues such as water resources and energy development, moving far beyond his previous emphasis on nature conservation and into the much more politically and economically sensitive realm of pollution prevention and resource use in urban areas. Still, two years later, when I visited with his staff in summer 2002, they appeared focused almost entirely on expanding their environmental education efforts rather than on undertaking new issues. An innovative program is Friends of Nature's environmental education van, which transports volunteers to China's poorer, interior provinces, where they live for several weeks at a time and use pictures and games to educate schoolchildren on the particular environmental challenges of the region. Friends of Nature is also using its relatively senior status and strong endowment to provide small grants—perhaps one to two thousand dollars—to help nascent NGOs get started throughout the country.

Yet Liang's greatest contribution to environmentalism in China may well be in his service as an activist above reproach. With his distinguished reform lineage, as well as his nonconfrontational approach to change, he is able to work within the system to push the boundaries of what the government considers acceptable activities for NGOs. Moreover, with high official and unofficial standing in government circles, he is able to serve an important role in conveying other environmentalists' concerns to government leaders. For example, one NGO based in Yunnan, the Green Plateau Institute, sent a letter to the National Minority Affairs Bureau via Liang, so that he could use his influence to press the government to stop government-sponsored groups from climbing the holy Meili Snow Mountain in Yunnan's northwestern region. Thanks to Liang's intervention, the treks were halted. Nonetheless, even Liang's Friends of Nature is not exempt from the government's strictures. In 2003, the government threatened not to renew the organization's registration as an NGO unless the secretary and a co-founder of the organization, Wang Lixiong, were dismissed from the organization. Wang had publicly come to the defense of two Tibetans convicted of separatist terrorist activities earlier in the year. Although Wang fought his dismissal, Friends of Nature decided that he should leave rather than endanger the organization's future.[53]

The Monkey and the Antelope

Liang also played a central role in two campaigns that galvanized environmental activists throughout China: the campaign to protect the Yunnan snub-nosed monkey, which was threatened by illegal logging in Deqin County, Yunnan, and the effort to protect the Tibetan antelope from poachers in the western reaches of China. These campaigns not only energized the environmental activists but also helped catalyze a nationwide environmental movement and establish a foundation for future campaigns.

In 1995 nature photographer Xi Zhinong, while employed by the Yunnan Forestry Department, had spent months documenting the ways of the near-extinct Yunnan snub-nosed monkey, which makes its home at the edge of the Tibetan plateau in Deqin, Yunnan.[54] The monkeys' diet consists almost exclusively of lichens from the

branches of the trees in the old-growth forests on mountains sepa-
rating the upper Yangtze and Mekong rivers,[55] and logging—both
legal and illegal—had devastated the monkeys' habitat. Soon after
completing his filming, Xi learned that the county government had
plans to sell extensive tracts of the forest to loggers. When Xi
protested to the provincial forestry department's leaders, they explained that the forest lay outside protected areas. Even the media,
many of which had become actively engaged in environmental investigative journalism, refused to become involved, concerned that
the issue was "too sensitive" because the Tibetan minority was involved.[56]

A friend of Xi's then contacted Tang Xiyang, describing Xi's work
and the plight of the monkeys. Tang followed up by calling Xi and
persuading him to write a letter to Song Jian, then the head of the
State Environmental Protection Commission, linking concern for
the nearly extinct monkeys to the larger threat of deforestation
along the upper Yangtze, noting that silt from the denuded slopes
could clog the Three Gorges Dam.[57] At the same time, Tang passed
the letter to Liang Congjie, who then passed copies of the letter to
the media. The letter was first published by the Associated Press
and quickly picked up by the Chinese media, generating significant
domestic and international attention. Concurrently, budding journalists and student environmental activists, Wen Bo and Yan Jun,
organized a group of more than two hundred students at the Beijing
Forestry University to view Xi's film at the end of 1995.

These letters and the exhibition generated a tremendous amount
of attention within various state agencies. As a result of this internal and external pressure, in April 1996, the government launched a
formal investigation. China Central Television's (CCTV) *Evening
News (Xinwen Lianbo)* followed up with a program on the logging,
did a feature story on Xi, and invited him to work on the popular
television program *Oriental Horizon.* Later, in summer 1996, Xi
helped Tang manage the first Green Camp in Deqin, leading over
thirty students to the forest and adding to the publicity. Finally, the
State Council instructed the local government to stop the illegal
logging and promised to compensate the loggers for a three-year period at a rate of eight million yuan ($970,000) annually.

In 1998, however, Xi was tipped off that the logging was continuing. CCTV's investigative program *Focus* went undercover and filmed the logging, following up with an interview with the Fuxianzhang County head. By chance, the program aired the evening before the devastating Yangtze River floods. Premier Zhu Rongji then stepped in to take control of the situation personally, phoning the director of Yunnan's forestry department to insist that action be taken. As a result, the Yunnan Forestry Department's vice director held a meeting at which the head of Deqin County had to "confess his mistakes," and the deputy-head of the county was fired.[58]

Although sixteen high-level officials signed onto the campaign, the local government has not effectively halted the logging, likely because the economic interests are too strong.[59] Some officials have threatened that they could "make Xi disappear,"[60] and he was fired from his job at the Yunnan Forestry Bureau. Liang Congjie believes that the government and NGOs will have to undertake greater grassroots education in Yunnan to educate the public and the officials for the protection effort to succeed.[61]

While the campaign may have attained only limited success in preventing deforestation and protecting the monkeys, it transformed the landscape of environmental protection in China, galvanized the environmental protection movement, and gave the activists a sense that they could indeed accomplish something through their efforts. It was the first time China's environmentalists had coordinated their activities and affected policy at the most senior levels of the Chinese government.

Several of those involved were inspired to become even more involved in the NGO movement. For example, Xi and his wife, Shi Lihong, returned to Deqin in 1999 to try to register their own NGO. Shi was already a veteran activist in the world of international environmental NGOs. Originally a reporter with *China Daily*, Shi had viewed a documentary on women environmental activists produced by Liao Xiaoyi and was inspired to take action. In 1996, she, along with Wen Bo and several other young activists, attended the Deqin Green Camp sponsored by Tang Xiyang. Shi then joined Friends of Nature and Global Village of Beijing as a volunteer, and began working part-time at the World Wildlife Fund (WWF) as an environmental education assistant in March 1997 and then as a full-time communications specialist in October 1999.

Yet, despite their excellent credentials, when Shi and Xi applied for NGO status, they were turned down by the local Party secretary, likely because of Xi's campaign to save the monkeys.[62] At the provincial level, however, top officials were favorably disposed to their work, and their application was approved. Finally in March 2000, they registered as the Institute of Ecological Conservation and Development (now called the Green Plateau Institute)[63] under the Xicheng District Bureau of Commerce and Industry in Beijing. They focused their work primarily on biodiversity and soil erosion, although much of the emphasis was on "community based natural resource conservation and development."[64] They undertook a variety of projects such as enabling local women in Deqin to sell their carpets, working to enlarge a small regional hydropower station, assisting environmental architects interested in improving the interior of Tibetan homes to use less wood, and training teachers. They also continued their investigative efforts. Reporting on the government's efforts to clean up Dianchi Lake, Shi Lihong stated that despite billions of dollars of investment, the water quality in the lake had actually declined.[65]

The energy and personal sacrifice required to serve as a leader of China's environmental movement, however, takes its toll on even the most committed. I met up with Shi while she was attending the 2002 World Economic Forum in New York as one of the "World's 100 Future Leaders." She told me that she and her husband had left Yunnan and returned to Beijing. Despite all their hard work in Yunnan, they were not native to the area, and decided to leave their work in the hands of a local environmental activist that they had trained. Back in Beijing, they had already established an independent documentary studio, Wild China, which concentrated on producing television movies on rare species in China.[66] Six months later in June in Beijing, I met with Shi. Normally dynamic and articulate, Shi seemed exhausted and concerned about leading a life that placed her constantly under the spotlight. Liao Xiaoyi had earlier counseled Shi that life as one of China's top environmental activists necessitated a substantial loss of privacy and sacrifice of one's personal life. Perhaps bearing such advice in mind, Shi decided to accept a long-standing invitation from a well-known China scholar, Orville Schell, to take a year sabbatical in the United States at the University of California, Berkeley.

The cause of the Tibetan antelope provided a second rallying cry for China's nascent environmental movement. Over the past several decades, the Tibetan antelope has become increasingly endangered as a result of the international trade in shahtoosh, the wool from the neck of the antelope, which is coveted for its exceedingly fine quality. The link between the decline in the numbers of antelope and the shahtoosh trade was originally established by noted American environmentalist George Schaller, who has spent years in China documenting and working to preserve China's rare species. Although the Tibetan antelope has been under the protection of the Convention on International Trade in Endangered Species (CITES) since 1979, their numbers have declined from almost 1,000,000 at the turn of the last century to under 75,000 by the turn of this century.[67] If poaching continues at the current rate, the species will likely be extinct by 2020.[68]

The antelope range throughout Kekexili, Qinghai, as well as Xinjiang and Tibet, a vast, largely barren region of western China. The shahtoosh trade is driven by three sets of actors: the poachers; the producers of shawls made from shahtoosh, primarily based in Kashmir; and, most important, the very wealthy consumers in the West, where a shahtoosh shawl may retail for more than $10,000. Crackdowns have taken place in the United States and England, among other western countries. India, however, has remained impervious to entreaties from environmentalists to shut down its mills. Dr. Farooq Abdullah, until 2002 the chief minister of Kashmir, stated that "as long as I am the chief minister, shahtoosh will be sold in Kashmir." He argued that there was "no evidence of Tibetan antelope being reduced in number or their being shot to acquire wool for shahtoosh."[69]

Within China, a battle between poachers and the Wild Yak Brigade, an unofficial group that patrols the area in search of poachers, has emerged.[70] The brigade was first formed in Zhiduo County, of which Kekexili is a part, and initially headed by Gisang Sonam Dorje, a local official. When he was killed by poachers in 1994, his brother-in-law, Zhawa Dorje, assumed his position. In 1998, Zhawa Dorje was found dead from a gunshot wound in his home in Golmud, Qinghai; it was ruled a suicide, but other explanations including murder and a family argument have also been advanced. At that same time, Liang Congjie, Xi Zhinong, and Yang Xin of Green River Network decided to become involved in the effort to protect the antelope.

In 1997, Xi traveled to Tibet with Yang Xin to profile Yang for *Oriental Horizon* and spent two weeks on patrol with the Wild Yak brigade. Upon his return, Xi persuaded Shi Lihong to involve the World Wildlife Fund. Shi's boss, Lu Zhi, then spoke to Liang Congjie and Grace Ge, the International Fund for Animal Welfare (IFAW) representative in Beijing, about what NGOs could do to help. Shi began collecting information on the challenge, eventually contacting the Convention on International Trade in Endangered Species office in Geneva, which agreed to participate in a meeting sponsored by WWF along with IFAW in Geneva.

In July 1998, Friends of Nature and a young journalist from the forestry bureau's newspaper *Green Times*, Hu Kanping, invited Zhawa Dorje to give a series of lectures in Beijing. The lectures inspired the public and mobilized support behind the Wild Yak Brigade; just one month later, however, Zhawa Dorje was killed.

Despite this tragedy, the Beijing-based NGOs continued to press forward. Yang Xin, who had been a close friend of Gisang Sonam Dorje, began to work to create a vast preserve in the region that would house facilities for climate change research, as well as five ranger stations to monitor the area and combat poachers. Simultaneously, Liang Congjie and Friends of Nature raised money to buy two Beijing Army jeeps for the brigade.[71] Liang also sent a report about the Tibetan antelope to NEPA and the Ministry of Forestry, calling for more funds, weapons, and manpower. Most important, Liang recognized that if the poachers were to be stopped, all three provinces and regions that were home to the antelope would have to crack down simultaneously.

The NGO effort also received assistance from the media. In spring 1999, after a year of effort, Xi Zhinong was able to persuade budding young television producer Hu Jingcao to produce a feature on the Tibetan antelope. After officials were persuaded to cooperate in a crackdown in April 1999, more than twelve groups of poachers were arrested. But by June, the poachers were back in action.

In early August, the Qinghai government dealt a serious blow to the antipoaching efforts by stating that it wanted to replace the Wild Yak Brigade with an "official" nature reserve administration. Liang Congjie and other conservationists, along with the Wild Yak Brigade itself, strongly opposed this shift, in part because training and developing such an official presence would likely take upwards of a year but also, one might presume, because the opportunities for corruption would be great. Despite personal lobbying of the NGOs by Qinghai officials, the NGOs resisted the change.

Still, some NGO leaders had their own reservations about the activities of the Wild Yak Brigade. The group had been illegally charging local residents to harvest worms in the area. In addition, there were accusations within the group of misappropriation of funds. However, after so much time and effort had been invested, the NGO community in China was reluctant to let the Wild Yak Brigade disintegrate. As one NGO leader commented, "The Wild Yak Brigade has become a sort of flag for the Chinese environmental community. No one wants to let it fall, but now many people are afraid to be associated with the brigade because of their behavior."[72] By 2001, however, Qinghai officials had succeeded in disbanding the brigade and had organized an official effort. At least one NGO leader believes that this official effort is now succeeding.

The Tibetan antelope and Yunnan snub-nosed monkeys have served as causes célèbres for the Chinese NGO environmental movement, both solidifying a nationwide network and inspiring new activists to take part. They also exemplify the reach of the Beijing-based NGOs and their capacity to alert the national leadership to environmental problems and work in support of national environmental policy at the local level, when local actors are ineffectual. Yet if China's environmental movement is to thrive, it must broaden the range of activities it supports and incorporate more regionally based NGOs to sustain the work over the long run. This has proved a difficult task. Outside Beijing, the nongovernmental movement is less politically well-connected, possesses fewer ties to the international NGO community, and is more often subject to the vagaries of local politics.

Grassroots Environmentalism in the Hinterlands

Many regional NGOs have emerged in cities and towns rich in biodiversity, such as Kunming, Yunnan. In several cases, the government has been effective in constraining the development of NGOs through its registration requirements. However, even with limited human and financial capital, these NGOs have proved remarkably adept at sustaining their work; in fact, there are some striking examples of environmental activism that reach beyond what has been attempted by Beijing activists, which may well serve as models for future environmental activism in China.

Yang Xin's Green River is perhaps the most successful of the regional NGOs. Yang, an accountant in his late thirties, has established an NGO with an annual budget of 440,000 yuan ($55,000).[73] He focuses his work on environmental conservation in the headwater of the Yangtze River and has published a well-received book, *The Source of the Yangtze*, which has already had significant U.S. sales and will bring in additional revenue for the organization. He is also quite organizationally savvy, spending part of each year in Beijing and maintaining close links to Beijing-based NGOs such as Friends of Nature and Global Village of Beijing. In addition, Yang has developed a website for Green River.

In spring 2000, Yang organized scientists, reporters, and government officials to travel to the source of the Yangtze to install a stone monument with an inscription from Jiang Zemin. Yang's most significant work, however, as noted earlier, has been the establishment of the nature protection station in the headwater region of the Yangtze, developed in part as a response to the poaching of the Tibetan antelope and in part to encourage ongoing research and environmental monitoring in the area.

Yang is successful because of both his personal abilities and his good connections with local government officials. He had no difficulty registering as an NGO in 1998, although the Sichuan EPB warned him to stay clear of politics.

Other NGOs, like the Green Volunteer League of Chongqing and Action for Green in Kunming, have encountered serious difficulties in sustaining their environmental work. Green Volunteer League was founded by Tian Dasheng, a professor of German at Chongqing University, who developed a love of nature through Russian author Ivan Turgenev, Tang poetry, and the German Romantics and was inspired to take action by the leftist Green movements in Sweden and Germany during the 1980s. In 1996, after Tian learned about Friends of Nature from a television broadcast, he joined the organization and participated in a tree-planting campaign in Inner Mongolia. When he returned to Chongqing University, he helped establish a student group called *Luse Jiayuan* (Green Homeland).

Wu Dengming, co-founder of Green Volunteer League, was formerly a People's Liberation Army (PLA) officer. He views his involvement in environmental protection as a means of repaying the debt he feels for the wildlife he helped destroy as the leader of the hunting team for his PLA unit during the Great Leap Forward in the 1950s. As a steel worker, he also worried that China would end up as polluted as western industrialized countries. He became an activist after he met Tian, who inspired him with stories of environmental protection.

In establishing Green Volunteer League, Tian and Wu faced several obstacles. Tian originally planned to establish a branch of Friends of Nature; however, he was instructed that branches were not permitted. Indeed, he was told by the local government that since there was already an NGO in the region, he wouldn't be permitted to establish one at all. Together, Wu and Tian then approached the leaders of what had become a defunct organization, the Green Volunteer League of Chongqing, took it over, and adopted their registration. Wu and Tian have tried to improve their position by establishing ties to other NGOs. Both Wu and Tian are members of Friends of Nature, and they participated in Liao Xiaoyi's Earth Day activity in March 2000 in Beijing. As a result of that activity, they also began to work with Yang Xin.

With about seventy members in their organization, Tian and Wu have pursued a bold approach to environmental protection. They went undercover to expose illegal logging activities, in one case posing as a businessman and geologist, gaining the confidence of the loggers. Their investigation and subsequent report led to the head of the village being sentenced to three years in jail, although in the end it turned out to be a relatively relaxed form of house arrest. With extensive media coverage of such exploits, Wu and Tian became celebrities of sorts.[74]

Duan Changchun has also achieved some significant success. With assistance from Xi Zhinong, Duan established Action for Green in 1999 and one year later mobilized students for Earth Day to meet and talk with people in downtown Kunming about how to practice a green lifestyle. They reached over 10,000 people, and 3,000 Kunming residents signed "green lifestyle" commitment cards. In addition to his work with Action for Green, Duan also advises the Green Environment Organization, a student organization with about 400 members that holds an annual green summer camp.

Despite these successes, however, both NGOs have had a difficult time surviving. Finances are a key sticking point. Wu and Tian received a $10,000 grant from the Civil Society program of the Canadian International Development Agency (CIDA).[75] Still they struggled to come up with the money to meet the government's 200,000 yuan ($25,000) funding requirement and were forced to use their retirement funds to pay for the administrative costs of the Green Volunteer League. Duan's situation was even worse. Although he attempted to register his organization as an NGO in 1996, he had no access to the funds needed. It wasn't until 1999 that Action for Green was approved, after receiving a grant from the Global Greengrants Fund, an international NGO that provides small grants to fund grassroots activities in impoverished and environmentally degraded areas throughout the world.

Duan ran into several additional roadblocks in his efforts to establish an NGO. First, he was compelled to change the name from Green Action to Action for Green because the former was perceived as politically threatening. Second, he did not have a full-time staffer or his own office space. In the end, the group was able to register as the Yunnan Environmental Science Society. Now, however, Duan's difficulty is that his organization, like the Green Plateau Institute, is classified as a second-tier one; thus, it cannot directly accept members. All new members of the group must become members of the Yunnan Environmental Science Society, which imposes scientific qualifications for membership. Since most of the associates of Action for Green are not environmental specialists but photographers, college students, and teachers, they don't have the credentials to join. Thus, although Duan's group has over seventy core members, officially, the group consists of only four people. Moreover, although the Yunnan Environmental Science Organization does not participate in Action for Green or provide it with any financial assistance, it must be notified of all events organized by Action for Green. As Duan states, "Their most important function is control."[76]

Nonetheless, Duan has demonstrated a willingness to push the political boundaries of environmental activism, however gently. In an unpublished paper, "The evolution of political affairs in response to eco-environmental issues," he touches on the continuing dispute over the wisdom of the Three Gorges Dam. He argues that environmental problems in China cannot be adequately addressed until the political situation is reformed:

> Like any issues in society, environmental issues may be well recognized and treated if "good directors" fortunately govern an area. But no one can guarantee a director is good enough to make reasonable decisions as to improve environmental quality. Only if knowing how to elect good directors and decide who are good enough to do well, the social mechanism can ensure increasing improvement of environment. Decisions as to environmental protection require making [decisions] on the base [sic] of well understanding of scientific background and direct public participation.[77]

Despite their difficulties, Tian, Wu, and Duan all remain committed environmental activists with a deep philosophical belief in the need for environmental protection and social change. Thus, in several ways, both environmental and political, the NGOs focused on conservation issues tend to reflect the ideals put forth by Tang

Xiyang and Dai Qing, calling for greater official transparency and accountability, albeit in less directly oppositional terms. While many of the NGOs outside Beijing suffer from lack of funding, staff, office space, and other politically induced constraints, they continue to pursue their work. At the same time, they are beginning to establish linkages with NGOs more directly interested in urban renewal, such as Liao Xiaoyi's Global Village of Beijing, and thus broadening the range of their environmental activities.

Beyond Conservation: Cleaning Up the Cities

Like Liang Congjie, Liao Xiaoyi (or Sheri Liao as she is known abroad), has become internationally renowned for her work on environmental protection. In 1996, she established the Global Village of Beijing, with the purpose of improving the urban environment and advancing environmental education through television programming.

Originally a researcher at the Chinese Academy of Social Sciences, Liao traveled to the United States to serve as a policy fellow at North Carolina State's Center for International Environmental Politics. She began to study the role of environmental NGOs and produced a documentary, *Daughters of the Earth,* which featured the work of forty female environmental activists. Inspired, she returned to China and established the Global Village of Beijing, whose main purpose was to raise public awareness of the environment in China through television programs.

Liao's work differs from that of Liang and other conservation-oriented environmentalists in several respects. Unlike Liang, Liao has registered her organization as a business. When she initially attempted to register an NGO with the Civil Affairs Ministry in 1995, the two-year moratorium was in effect. In addition, she was told that she would have to solicit permission from her supervising agency prior to undertaking any new activities or holding meetings. By registering as a business, Liao avoided these difficulties and has maintained a greater level of independence, although, as a business, she has to pay taxes.[78]

Liao's organization also differs from most Chinese NGOs in focusing predominantly on urban issues such as community development. With government support, she established a pilot recycling project in Xuanwu district in Beijing that has now been expanded to other districts. She is also working with Xuanwu to make it a "green community." Eventually, Xuanwu will employ a wide range of efficiency technologies such as energy-saving lights, water-saving faucets, recycling bins, and so on. To accomplish this, Liao works closely with the local EPB as well as with other departments of the local government to ensure that they are vested in the process.

Liao has also supported the Dacheng Lane Neighborhood Committee in the Xicheng district of Beijing, a group of retired teachers who have banded together to clean up the local environment. Perhaps most impressive, Liao founded an environmental protection education and training center outside Beijing, where forty families practice a green lifestyle, which includes sorting garbage, reusing everyday articles, and not using chemical fertilizers.[79]

Liao's strategy, which has been an effective one, is to change government policy by example and through media attention. According to Liao, SEPA Administrator Xie Zhenhua himself has been persuaded of the efficacy of green communities and has indicated that fostering green communities should be one of the main tasks of SEPA. As of 2002, with the financial support of CIDA, Liao was hoping to develop green communities elsewhere in Beijing, as well as in Wuhan and Shanghai, where local officials have indicated interest.

Liao views Chinese environmentalism as a triangle, involving the government, private enterprise, and the nonprofit sector. In her view, the role of NGOs is to educate the public, help the government implement environmental protection policies, and encourage business to demonstrate more care for the environment.[80] She held a conference on sustainable consumption that involved senior Chinese officials, NGO leaders, officials from the United Nations Environment Programme and United Nations Development Program, and business leaders from companies such as BP and Shell.

In contrast to other environmental activists, Liao is resolutely apolitical in her public comments. In one interview, for example, she stated, "I don't appreciate extremist methods. I'm engaged in environmental protection and don't want to use it for political aims. This is my way, and my principle too."[81] She is extremely reluctant to become engaged in thorny political issues like the Three Gorges Dam. While she will state that she does not "like" the project, she adds that she does not want to "make trouble." Believing that her focus must remain on the positive work that she can do to change people's behavior with regard to environmental protection, she maintains an aggressively forward-looking approach. In one private chat, she told me that in response to pleas from Dai Qing to become involved in protesting Three Gorges, she argued, "We have a lot of things to do in China besides Three Gorges. The most important thing for us is to avoid other mistakes like Three Gorges."

At the same time, Liao has reached out to NGO leaders throughout China, holding an annual Earth Day celebration that attracts regional NGO leaders to Beijing as well as environmental activists from abroad. In 2002, twelve NGOs participated. Liao also took the lead in organizing an exhibition for the World Summit on Sustainable Development in Johannesburg, South Africa, during late August–early September 2002. While several NGOs attended the summit, many more were introduced to the international community through a World Bank–funded video. The video exemplifies the transformation in Chinese society since the 1992 UN Conference on Environment and Development, when China was embarrassed by its lack of nongovernmental participation.

Liao is a strong believer in the power of the media, and she was a pioneer in the area of media and the environment. Her television program, *Time for the Environment,* is broadcast weekly on nationwide CCTV, and she produces a number of specials. In 2002, with the support of the U.S. Department of Agriculture and CCTV, she completed a series of nine programs on model environmental medium-sized farms in the United States that she hopes to place on local television stations. Indeed, her work highlights the importance of the media as purveyors of information and as a foundation for future environmental NGO activism.

Not only Liao's work but also her administrative skills serve as a model for other NGOs. Liao has raised $500,000 for her operations, primarily from overseas sources, and maintains a full-time staff of seven in a small but bustling complex of offices in the Asian Games complex on the outskirts of Beijing. Her extensive knowledge of the media, government, and NGO communities both at home and abroad is unparalleled.

The Media and Environmental Advocacy

Spurred in part by Liao Xiaoyi's pathbreaking work, the media have become an essential element of environmental activism in China. In a 2001 public opinion poll, almost 79 percent of Chinese indicated that they learned about environmental protection-related issues primarily from television and radio.[82] Government publicity was a distant second at 42 percent.[83]

In recent years, many media personalities and journalists have assumed leadership roles in environmental education. Wang Yongchen, the effusive director of Green Earth Volunteers, was perhaps the first radio host to explore environmental issues, using her talk show to raise issues such as the Tibetan antelope. In 2003, Wang helped lead a campaign—successful to date—to halt the construction of the Yangliuhu Dam on the Min River in Sichuan Province, arguing that the dam could damage Dujiangyan, possibly the world's oldest functioning irrigation system. China's mainstream media criticized the dam heavily, and the Chinese public voiced its strong opposition on Internet websites. Wang argues that this is the first time that the Chinese people have had input into such an important project.[84]

Chinese television is also playing a critical role in environmental protection efforts by investigating and protesting environmental crimes. As one former CCTV anchorwoman stated, "Seeking justice for the public has evolved as an important function of the Chinese media. Such a function may be part of China's *guoqing*—unique national situation. It is unique because rule of law has not been fully established in the country."[85] One of the most popular such television programs is *Focus*. Modeled on the U.S. news program *60 Minutes*, *Focus* went on the air in 1994 as an outgrowth of another trial investigative program, *Oriental Horizon*. While only fifteen minutes long, *Focus* has become a phenomenon on Chinese television, drawing 200–250 million viewers and spawning a range of investigative efforts into environmental and other issues within the Chinese media.[86]

In a fascinating paper on the role of the media in China, Li Xiaoping, an executive producer of *Focus*, recounts how people wait in line outside the *Focus* studio to ask the reporters to investigate various cases of environmental or other types of official wrongdoing. She also notes that bureaucrats have developed a saying, "Avoid fires, avoid theft, avoid *Focus*."[87] The local media are also getting into the act. In the mid-1990s, a Shanghai restaurant owner finally paid his fines to the local EPB after the local television station reported on his pollution.[88]

The media also play a prominent role in enlightening top government officials about environmental problems. Former premier Zhu Rongji was especially prone to take action after learning of environmental problems via television programs; his campaigns against desertification and illegal logging in the late 1990s both stemmed initially from television reports.[89]

Grassroots environmentalism may also be spurred by television programming. For example, the media have played an important role in engendering a public groundswell of support for battery recycling in China.[90] In one case, Geng Haiying, a young doctor from Dalian, first became interested in the environment after viewing a television program on battery recycling. The program was produced by Hu Jingcao, a young woman who had become integrated into the environmental movement in Beijing through Liang Congjie and others, producing a number of environment-related programs on issues such as the Tibetan antelope, the Huai River, and deforestation.

Based on Hu's program, Geng suspected that local dumps in Dalian, which were filled with batteries, were leaking and poisoning the local water supply and produce. She investigated the issue further through Green Beijing, SEPA's website, and by reading about the solid waste problem. She then single-handedly began a program to recycle batteries in Dalian. The local EPB was supportive, but since there was no formal law on the books, it did not take an active interest in addressing the problem. However, when SEPA stated that battery recycling would be a priority, she extracted a promise from the Dalian EPB to collect batteries. She then approached department stores to see if they would be willing to serve as collection points for the batteries. While the department stores were initially suspicious of her motives, by April 2000, she had persuaded three of them to establish recycling bins. Her environmental activism was also supported by Wen Bo, a Dalian native, who helped her gain access to outlets such as radio programs for publicizing her work.

Geng's experience touches on one of the greatest difficulties that grassroots environmentalism faces: the failure of the state to address the environmental protection desires of the populace. In 2000, another crusader for battery recycling, Tian Guirong, had collected 30 tons of batteries from throughout Henan Province. However, she commented, "I collected them to help with environmental protection, but I didn't realize there would be nowhere to dispose of the batteries."[91] In Beijing, the EPB has established a hotline and "Useful Rubbish Recycling Center" to collect used batteries and other refuse, but its disposal site for batteries does not meet the state's own environmental standards, so the Center has had to hold on to the batteries. According to one official, the staff is so fearful of phone calls asking where batteries may be sent that it is suffering from "battery-phobia."[92]

Thus, even with media attention and strong public support, China's environmental protection bureaucracy falls short of meeting public needs. Indeed, the range of individuals in China undertaking private environmental activities, such as private nature reserves and garbage collection,[93] only points up further the weaknesses of the local EPBs.

Geng Haiying's experience, however, also suggests that the combination of an activist orientation and media attention can generate further action. In fact, many of the next generation of environmental activists have been trained as journalists in China, often with further study and fellowships at universities, think tanks, and

NGOs abroad, and they bring to environmental protection not only strong technical expertise but also facility in communicating their ideas to audiences both at home and abroad. They will be a potent force for environmental protection, some in service of Beijing, but many more in service of the broader public's interests.

The Next Generation of Activists

The future of the environmental movement in China will depend largely on the outlook of the next generation of environmental activists. There is reason to be optimistic. Tang Xiyang, Liang Congjie, and other NGO leaders are cultivating this group closely. This new generation has been heavily influenced not only by its interaction with senior NGO leaders but also by its international exchanges. The result is a highly skilled, articulate, and politically savvy set of environmentalists, among them Wen Bo, reporter Hu Jingcao, the co-founder of Green Plateau Institute Shi Lihong, and Hu Kanping, editor of *China Green Times*.

Preeminent among these young environmental leaders is Wen Bo. At twenty-five years of age, Wen was already a renowned environmentalist with excellent journalistic and environmental credentials. In many ways, Wen exemplifies the new environmental activists in China. Early on, he interned at *China Daily*, working for Shi Lihong, and then continued to write for several years for *China Environment News*. At the same time, he organized student environmental groups at universities throughout China, helping to link them into the China Green Students Forum, which now has a membership of 250 student groups. He also participated in Tang Xiyang's Green Camp, and has spent substantial time working with international environmental NGOs such as Environmental Defense, International Snow Leopard Trust, and International Rivers Network, which is the most prominent lobbying organization opposed to the Three Gorges Dam. In 2001, Wen assisted Greenpeace in establishing its first office on the mainland.

Wen advises many other budding young environmentalists, such as Li Li, who served as the coordinator for Green Students Forum, and Wang Luqing, who founded the Dalian Student Environmental Society while a student at Dalian University of Technology.

Hu Kanping, unlike Wen, has remained a journalist, focusing his reporting on issues related to forestry for the State Forestry Administration. A literature major in college, Hu came of age politically in 1986–1987 in Shanghai, joining in student demonstrations in support of the political reform leader Hu Yaobang. An idol for many Chinese youth during the late 1980s, Hu Yaobang was ultimately stripped of his position as general secretary of the Communist Party in January 1987, and his death in April 1989 triggered the Tiananmen demonstrations. Hu Kanping has spent his career within the forestry administration of the central government as a reporter for the agency's newspaper, *Chinese Forestry News*, which was renamed *Green Times* in 1997.

Both Wen Bo and Hu Kanping recognize the limitations of the media. In discussing *Green Times*, for example, Hu stated that while his reporters are permitted to report on local corruption, they cannot report on violence or personnel issues. Moreover, while the forestry administration will give free rein to reporters to publish articles on issues like pollution that are outside the bureau's jurisdiction, it may limit reporters' publication of articles on some areas under its jurisdiction.[94] However, Hu has also pushed the boundaries of the acceptable, attempting to make environmental reporting accessible to the general public. He and Wen, for example, first became friends in October 1995, when Hu attended a newspaper bazaar to peddle his independent newspaper *Green Weekend*. While not a commercial success, it attracted significant positive attention from the forestry officials who appreciated the lack of official pronouncements and the lively, informative writing style. They encouraged him to enliven the official newspaper with such articles and writing. As Hu became well-known for his work, he was later invited to join the board of Friends of Nature.

China Environment News, which is supported by SEPA, has seen its autonomy ebb and flow. Established in 1984, the newspaper at first published a wide range of articles on pollution, conservation, biodiversity, environmental technologies, and new laws or policies issued by the government. It also frequently published exposés of the corruption and violence that occasionally marred environmental protection efforts. Moreover, it was free to report on the activities of international environmental organizations such as Greenpeace, although the purpose of such reporting was not to raise awareness of environmental problems in China but rather to highlight opposition to the practices of advanced industrialized nations.[95] By the late 1980s, *China Environment News* was published three times a week, with a total circulation of approximately 500,000, and had an English language version widely available outside China.

During 1995–1996, coverage of domestic environmental issues increased almost 100 percent; in 1996, China's seventy newspapers carried 17,555 environment-related articles.[96] (By 2000, such articles totaled 47,000.) But government attempts to control environmental reporting continued. In August 1995, when five foreign Greenpeace activists held a protest in Tiananmen Square against nuclear testing in China, the Chinese government banned all reporting of Greenpeace activities and reduced the amount of reportage on international environmental NGO activity.

And, in 1998, when the Hong Kong–based *South China Morning Post* cited a *China Environment News* article that was critical of the Three Gorges Dam, it generated significant concern within SEPA. Later in the year, SEPA halted the English edition of *China Environment News,* reportedly for financial reasons.[97] Still, the paper in Chinese remains a useful source of information domestically and abroad for understanding Chinese efforts at environmental protection, and an online version of the newspaper provides readers abroad with easy access to a wide selection of the paper's articles.

Moreover, young television journalists still aggressively pursue cases of environmental wrongdoing, exploiting the public's appetite for such reporting. After producing a program that highlighted the Huai River Valley disaster and the government's response, Hu Jingcao returned to the River for a follow-up program, only to discover that the factories were evading the government's dictates. This forced the leadership to acknowledge the shortcomings in the campaign.

The orientation of this younger group of environmentalists is open and aggressive. For example, one young journalist with whom I spoke said he would like to see NGOs increasingly challenge central government policy, produce high-quality reports that will help increase the government's efficiency and effectiveness, act as lobbyists and pressure groups, and use actions like those of Greenpeace to dramatize environmental issues. Still, he remains cautious about voicing his opinions openly, preferring to push the boundaries through his actions.

Others are even more explicit in their demands for change:

China's transition to a market economy has broadened the base for civil society. At the same time, the government is still very powerful. I think that environment groups can develop under the current situation, but that in the end, environmental work may lead to greater democracy in China. In fact, environmentalism and democracy are related. Many NGO leaders are hesitant to say that we are related, but I believe that the NGO movements are creating democracy.[98]

Overall, the tone is one of wariness but also cautious optimism. This generation of environmentalists seems to understand that the very existence of the nongovernmental sector is tenuous, but it believes that time is on its side.

There is no real environmental movement because the government would be very unhappy if there were a movement. The Party has almost all the power and NGOs have no money, no power, and no social resources. The government wants NGOs to help but not to make trouble [*keyi bangmang bu neng tianluan*]. But I agree with Tang Xiyang. More and more media and NGO focus on environmental issues will lead to more democracy and greater freedom. The hope is that we will educate the young people about the problems we have and the need to do something about them. In twenty years, these young people will be the leaders of China. In this way, we will build-up a democratic country.[99]

Hong Kong NGOs: Future Partner or Future Model?

While the next generation of mainland-based environmental activists has been influenced by its strong ties to organizations abroad, as well as its close links to senior Chinese NGO officials, it is also looking to Hong Kong as a source of funding, organizational assistance, and collaboration on technical and other types of environmental studies.

The reversion of Hong Kong to the mainland on July 1, 1997, following ninety-nine years of British rule,[100] engendered much speculation concerning the flow of political and economic influence. Would Hong Kong become more like the mainland, or would the mainland eventually resemble Hong Kong? Under British rule, Hong Kong had developed a number of proud traditions, including a highly skilled civil service, freedoms of press, speech, and assembly, relatively low levels of corruption, and democratic procedures in some elections. Hong Kong also developed a small but active environmental NGO movement, including nature associations, broadly supported by international NGOs. These NGOs appeal to a wide range of citizens, from nuclear power activists to bird watchers.

Despite coming from a very different political tradition, the majority of environmental actors in Hong Kong operate much like their counterparts in Beijing, allying with the Environmental Protection Department in opposition to other government or nongovernment actors, and focusing on environmental education and typically nonthreatening activities such as recycling campaigns and green agriculture.[101]

But a few Hong Kong NGOs adopt a more confrontational approach, lobbying against projects such as the airport in the Chek Lap Kok Island area, a golf course, and the PRC's Daya Bay nuclear plant.[102] Some of those most actively engaged in confrontational activities are branch organizations of international NGOs, such as Friends of the Earth, the World Wildlife Fund for Nature, and Greenpeace. For example, Friends of the Earth, Hong Kong, and Greenpeace are actively working to stop the construction of a nuclear power plant on the mainland, in Yangjiang, Guangdong, by the

Guangdong Nuclear Power Group. Construction was set to begin in 2002 and to be completed by 2015, but has yet to begin. The plant would be the third in Guangdong after Daya Bay and nearby Lingao, a twin reactor station that became fully operational in January 2003. Plato Yip Kwong-to, former assistant director of Friends of the Earth, Hong Kong, and a significant figure in Hong Kong's environmental movement, commented, "The Chinese say it's safe, but if something happens we only have two hours to evacuate Hong Kong. Just like the incident in Chernobyl, you never know when the next one could be coming."[103] Director of Greenpeace in Hong Kong, Ho Wai-chi, adds, "It's the wrong way to go. The Yangjiang plant will only create nuclear waste when they should be relying on other sources like wind and solar power."[104]

Despite the success they have achieved, environmental NGOs in Hong Kong face challenges similar to those on the mainland. The Hong Kong populace is difficult to mobilize on behalf of environmental issues, with the exception of concerns such as Daya Bay, which are perceived as life-threatening. Fund-raising is problematic given the overall apathy of many businesspeople about environmental protection. And according to one NGO leader, many Hong Kong NGOs have suffered because of the exodus of foreigners after 1997.[105]

Moreover, many in Hong Kong's business elite continue to stress development at the expense of the environment. Sir Gordon Wu Ying-sheung, chairman of the Hopewell Group, which is responsible for significant infrastructure development in both Hong Kong and on the mainland, condemns NGOs as stymieing economic development:

> If Hong Kong is to be run by consultants and environmental NGOs, then Hong Kong is dead. . . . A lot of people will be out of work. But never mind. They will collect their welfare cheques and have all the time in the world to watch the birds. That's the future of Hong Kong. . . . This environmental thing will eventually topple Hong Kong's development. You must make clear what environmental protection is. . . . The most endangered species is called Homo sapiens. . . . Keeping the air and water clean is fine. But it

shouldn't be a question of just a few tree frogs. . . . Say you build a road, and there are tree frogs, birds or dolphins nearby. They will go away and return. Look at the map. See how big the world is? They can find some place to settle. But we humans are different. We can't fly. Without development, there will be no jobs.[106]

The challenges of Hong Kong NGOs notwithstanding, they are shaping the practices of their mainland counterparts in important ways, providing detailed environmental information and crucial financial support. For example, Friends of Earth, Hong Kong, funded a two-year SEPA study published in 1998 on public opinion and the environment. They also established an annual Earth Award in September 1996 to acknowledge the contribution of China's journalists and educators to the environment. According to one mainland activist, this nomination process has created a network of lesser-known activists whom the group supports.[107] In June 2000, a delegation of Hong Kong green groups traveled to Beijing to present its concerns about cross-border pollution to members of the National People's Congress.[108] The Conservancy Association in Hong Kong also supports some environmental activists on the mainland, including Yang Xin, Wang Yongchen, and some lesser-known environmentalists in Sichuan and Yunnan provinces.

For the next generation of environmental leaders on the mainland, some of the Hong Kong NGOs also provide a nearby model for environmental information gathering and lobbying techniques. Some mainland environmental activists told me that the Hong Kong NGOs are often more helpful than those from western countries because they understand China, the nature of the political system, and the language. They also produce their materials in Chinese, making them especially valuable for those mainland NGO leaders who do not speak English. The growing exchanges between the two groups will likely provide some foundation for future cooperative NGO activity on the mainland. Bringing pressure to bear on Hong Kong-based businesses that are substantial polluters on the mainland, for example, would provide a natural venue for joint Hong Kong–mainland NGO efforts.

The Future of the Environmental Movement in China

In a China characterized by constantly shifting political winds, environmental NGOs have moved rapidly into the political space opened to them during the early to mid-1990s. They have had a significant impact in raising environmental consciousness among the Chinese people through Earth Day information campaigns and a barrage of television programs devoted to environmental issues. They have exposed inadequate local implementation of national environmental protection laws, especially with regard to species protection and deforestation. And they have begun to work in urban areas to employ environmentally sound technologies and practices to improve both the environment and people's lives. In these ways, they have served the interests of the Chinese people, their own concerns, and the needs of the government in Beijing.

Understandably, the NGOs have avoided open conflict with the central leadership in Beijing. Now, however, some environmental activists such as Liang Congjie and Tang Xiyang have begun to prod Beijing directly, opening the door to the possibility of more direct criticism and lobbying by the next generation of environmental leaders. We can expect this next generation to be bolder. They possess the full complement of skills necessary to organize effectively: technical expertise on the environment, strong backgrounds in journalism and media, and extensive ties to environmental activists both throughout China and abroad.

Why does Beijing tolerate the increasingly activist environmental NGOs? For one thing, they provide an inexpensive mechanism for monitoring local pollution efforts and educating the public on environmental protection. While SEPA has not established any extensive linkages or programs with environmental NGOs, their interests are frequently allied, and Xie Zhenhua has offered direct support for programs such as green communities that NGOs have initiated. NGOs also offer a degree of political cover to an otherwise authoritarian government, signaling to the world that China does tolerate independent societal organization.

Veteran NGO leaders in China are sensitive to the possibility that the government will use them for propaganda purposes. For example, in its bid to host the 2008 Olympics, the government solicited the support of Friends of Nature, Global Village Beijing, and twenty other environmental groups for the "Action Plan for Green Olympics." As Liang Congjie and Sheri Liao both commented at the time, the NGOs agreed to lend their support in hopes of realizing long-term environmental benefits. Liao stated, "Whether Beijing finally wins the 2008 bid or not, the Action Plan for Green Olympics will be of great significance to Beijing's sustainable development."[109]

Some positive change may indeed be forthcoming. Newspaper articles in the lead-up to the Olympics decision mentioned the possibility that Beijing's largest polluter, the Capital Iron and Steel Company, would have to be moved if the city won the bid, and in January 2003, Chinese officials announced a $341 million effort to relocate 75 percent of the company to a rural part of Qianan city in Hebei Province.[110] Such a move had been proposed five years earlier, in July 1996, when Liang's Friends of Nature held a forum on the pollution generated by the enterprise and issued a report suggesting where the company might be relocated.[111] In the wake of China's Olympic victory, however, Liang evidenced more concern than elation, noting, "My greatest worry is that the committee will focus on making Beijing into a showcase city with water-wasting stretches of grass."[112]

The government's support of NGOs remains qualified. The range of restrictions it has placed on NGOs regarding registration, funding, staffing, and location have constrained the number of NGOs as well as the range of their activities. If the government is committed to an active and engaged public effort to protect the environment, especially as a means of overcoming the state's decreasing capacity to meet the country's environmental protection needs, it will need to relax these restrictions and free NGOs to flourish.

But this is unlikely in the near future, unless there is a significant change in the political outlook of the top party leadership. While environmental activists, with the exception of Dai Qing, have been wary of pushing too hard or too fast for political reform as a means of enhancing environmental protection, they have not shied away from articulating similar political ideals. The Communist Party leaders recognize that, in the context of the rapid socioeconomic changes occurring in China today, the potential for environmental NGOs and other social groups to press for greater political change, as in the cases of some Eastern European countries, former Soviet republics, and Asian neighbors, poses a real danger to their authority and leadership.

CHAPTER SIX

THE DEVIL AT THE DOORSTEP

The Chinese leadership has embraced the international community as an essential component of its long-term strategy to improve China's environment, welcoming cooperation not only with other countries and international governmental organizations, such as the United Nations (UN), the World Bank, and the Asian Development Bank (ADB), but also with multinationals and international nongovernmental organizations (NGOs).

At the same time, China's economic reforms and its integration into the global economy have opened the door to new policy approaches and technological possibilities in environmental protection. Attributes valued by the market such as efficiency, transparency, rule of law, and managerial expertise have begun to permeate environmental protection in China. Moreover, China's participation in trade-based regimes such as the Asia-Pacific Economic Cooperation (APEC) and the World Trade Organization (WTO) now offers opportunities to reinforce the integration of environmental protection objectives into the process of economic development. Over time, the transformation of China's economy has the potential to redefine radically China's approach to environmental protection.

But significant impediments to international environmental co-operation remain. Much of the infrastructure for effective policy or technology implementation is not in place. Inconsistent enforcement of environmental laws, lack of transparency, poor management, and weak incentives for environmental protection all curtail the viability of many of the most desired technologies and policy tools. China's continued emphasis on economic development means that the leadership may also ignore pleas from members of the international community for more proactive policies on regional and global environmental issues. Moreover, the international community at times ignores China's real needs, focusing instead on technologies that may not be appropriate for China's current level of economic development and capacity and commitment to environmental protection.

China Meets the World

As we've seen, the 1972 United Nations Conference on the Human Environment (UNCHE) marked China's first foray into the arena of international environmental politics. China's participation raised the profile of environmental issues on the agenda of China's leaders and contributed to the establishment of the country's first formal environmental protection apparatus.[1] The UNCHE also launched China into a new world of global environmental concerns. Through the UN, the international community soon engaged China in battling trade in endangered species, marine dumping, trade in tropical timber, and the degradation of sites of natural and cultural value.

As the international community impressed on China's leadership the importance of participating in the full range of environmental treaties and regimes, China responded by establishing a leadership group for each treaty, consisting of the relevant ministries and agencies under the auspices of the State Council, to study the issues and develop recommendations. This process created small communities of experts, including many who had been forced to give up scientific or intellectual pursuits during Mao Zedong's reign, and who formed the basis of an emerging scientific and environmental elite.

In some cases, for example marine dumping, the Chinese were not significant contributors to the problem. In acceding to the London Dumping Convention, therefore, Chinese leaders' motives were primarily political: to rejoin the international community, to prevent Taiwan from asserting itself as the representative of China in international bodies, and to develop a coterie of trained environmental and scientific experts who could transmit relevant knowledge to the Chinese leadership to help with domestic environmental problems.[2]

But in order to sign on to any treaty, China had to develop the capacity to implement it. This required the drafting of new laws, the training of experts to understand and meet the demands of the treaty as well as to implement it, and coordination among as many as ten or more bureaucracies, each of which controlled a different aspect of the treaty.

This preaccession process is often long and complex. In the case of the London Convention against Ocean Dumping, the Chinese signed on in 1985, over a decade after the convention had been established internationally, first ensuring that their domestic laws and capabilities were reasonably in keeping with the demands of the international agreement. Before signing on, for example, China passed its own Marine Environmental Protection Law modeled on the convention. The long lead time also permitted the relevant Chinese ministries to study the domestic ramifications of the issue, jockey among themselves for the lead position in the international negotiations, and balance their competing interests to arrive at the formal Chinese negotiation position.

During the course of the negotiations, the Chinese representatives maintained a low profile, studying the technical points of the proposed treaty and developing expertise by participating in training seminars and technical development efforts. The various UN secretariats that oversaw the international environmental agreements typically offered extensive educational programs and some technology for the expert communities in developing states. Once China became party to the London Dumping Convention, China's State Oceanographic Administration took advantage of training courses on monitoring ocean dumping and developing an adequate legal system offered by the UN Environment Programme and the International Marine Organization.

Others in the international community also helped the Chinese develop the capacity to fulfill their treaty commitments. For example, several years after China acceded to the World Heritage Convention in 1985, the World Bank and the Getty Museum in Los Angeles together assisted the Chinese in restoring and protecting the Yungang Grottoes and the Mogao Grottoes of Dunhuang—ancient Buddhist sites that contain spectacular rock carvings, paintings, and painted clay figures from as early as the fourth century. They coordinated efforts to reduce sandstorms in the area and to conduct environmental monitoring inside the caves, and the Getty supplied instruments and trained workers. In turn, China exported some of its newly trained experts to Cambodia to contribute to the international effort to protect Angkor, the site of Khmer empire ruins, including the famous temple of Angkor Wat.

China's 1981 accession to the Convention on International Trade in Endangered Species was similarly aided by the World Wildlife Fund (WWF), the International Union for the Conservation of Nature (IUCN), and the U.S. Department of Interior, which helped with the training of Chinese officials, database development, and wildlife studies. The WWF and IUCN even worked with Chinese officials in an undercover sting operation to catch those who trade in tiger bone for use in Chinese traditional medicine, a practice that has contributed to rapid decline of the South China and Siberian tiger populations.[3]

During the 1980s, many of China's domestic environment and resource issues captured the attention of international actors. For example, the World Bank initiated projects on fisheries, rural water supply, and the development of natural gas, while the WWF became deeply involved in species protection in China, including a multiyear effort to protect the rapidly declining number of China's famed pandas in Sichuan Province.[4]

Raising the Stakes: Ozone Depletion and Climate Change

As China gradually developed the scientific and technical expertise as well as the practices to support its full participation in international environmental affairs, growing world awareness of ozone depletion and climate change in the late 1980s and early 1990s placed Chinese environment and development policies under intense international scrutiny and raised issues of environmental protection to the highest level on the agenda of the Chinese leadership.

Ozone depletion and global climate change pose potentially devastating health, environmental, and economic challenges to the countries of the world, up to and including, in the case of climate change, the destruction of some small island states because of sea level rise. These global environmental phenomena also differ substantially from other environmental problems because the treaties designed to respond to them entail substantial financial and human resource commitments.

Before the 1980s, China's contribution to these problems was inconsequential. Both global climate change and ozone depletion result primarily from industrial processes related to development rather than from natural phenomena. China's relatively late industrial development meant that compared to Europe and the United States, the country bore little responsibility for the current global environmental state. However, owing to China's sheer size and the rapidity and scale of its industrial development throughout the 1980s, China quickly became one of the chief contributors to both problems. In 1986, China's consumption of ozone-depleting substances was roughly 3 percent of the world total; by 2001, it had become the chief contributor to ozone depletion and remained so in 2003. Similarly, during the late 1980s, China ranked third in total contribution to global climate change; by the mid-1990s, China ranked second after the United States.

Participating in the treaty negotiations for the London Amendments to the Montreal Protocol on Substances That Deplete the Ozone Layer and the Framework Convention on Climate Change dramatically raised the stakes for China's commitment to global environmental problems. When negotiating the London Amendments, the government established a leading group under the auspices of the National Environmental Protection Agency (NEPA) to evaluate the costs and benefits of signing the protocol. This group concluded that there were three reasons China should sign the agreement. The most important was market-based—the treaty's potential sanctioning mechanism. China was interested in becoming a major exporter of light industrial products, such as refrigerators, that use ozone-depleting substances (ODS). The protocol, however, forbade signatories from purchasing such products and sanctioned those countries that traded in them. Second, some within the Chinese leadership believed that, as a member of the international community, China should contribute to the resolution of ozone depletion and that it would enhance China's image to sign. Finally, according to one member of the working group, strong scientific evidence and interactions with the international scientific community were also key reasons for supporting accession.[5]

Yet in the end what drove China's negotiating position were the financial implications of joining the Montreal Protocol. Along with India, China refused to sign the accord until significant financial support and transfers of technology from the international community were promised. Once the international community agreed to establish a special multilateral fund for this purpose, China ratified the treaty in 1991.

A similar pattern emerged with regard to global climate change. Because of the complexity of the science inherent in the problem of climate change, the formal negotiations were preceded by lengthy scientific discussions involving Chinese scientists, leading to unprecedented consultation between domestic and international expert communities. This convention also demanded research and data collection by the experts from each country. In China, this effort prompted the establishment of an extensive coordinating bureaucracy, whose job it was to collect the necessary data for the scientific discussions. By 1989, one year after scientific negotiations began, China had organized a climate change research program encompassing forty projects and involving about twenty ministries and five hundred experts.[6]

The international community was instrumental in funding this scientific research. The World Bank, the Asian Development Bank, the UN Environment Programme (UNEP), the UN Development Programme (UNDP), Japan, and the United States all provided monitoring equipment for greenhouse gas emissions, shared computer modeling techniques, offered technological assistance in developing response measures, and trained Chinese environmental officials. For example, the State Planning Commission's (SPC) Energy Research Institute became a focal point for international work on energy and climate change (despite having no history of such research) when an expert from the U.S. Department of Energy Lab, Pacific Northwest National Laboratory (managed by the nonprofit corporation Battelle), identified a few talented economists from the Institute and trained them in the United States. China eventually used a variant of one of the lab's models for estimating emissions of carbon dioxide in preparation for the political negotiations,[7] and one of the SPC's U.S.-trained experts later served as a member of China's climate change negotiating team. Several Chinese members of the various international climate change working groups emerged from their participation with a new appreciation of the importance of China's contributing to the international climate change effort.[8]

Yet as the negotiations moved from the scientific to the political, the influence of the international community diminished. Within China, the lead agencies were no longer the more proactive scientific, technical, and environmental ministries but rather the Ministry of Foreign Affairs and the SPC. Concerns over economic development and sovereignty now dominated the domestic debate, overshadowing the potential benefits such as access to technology, environmental capacity building, and international goodwill.

The majority of the Chinese leadership believed that the advanced industrialized countries should restructure and reform their own wasteful practices and not simply look to the developing countries as a cheap alternative. As one science official commented, "The policy making on climate change depends on social issues, not science."[9] In addition, the potential economic costs of taking action on climate change were far greater than those involved in reducing ozone depletion. A meaningful Chinese response to global climate change would require a significant reorientation of the Chinese energy industry and substantial investment in new energy efficiency technologies.

Therefore, in the end, despite the wide-ranging impact of the international community on the development of domestic Chinese institutions, the establishment of a scientific community, and a flow of funds to China for scientific research, China's stance on climate change was regressive. China refused to consider any targets or timetables for reducing its greenhouse gas emissions. It was unwilling even to permit other countries to undertake joint implementation activities, such as reforestation, within China in order to fulfill their obligations within the treaty.[10]

Subsequently, the international community and proactive domestic actors continued to press the Chinese government to adopt a more flexible position in its response to the challenge of climate change. During the late 1990s and into 2000–2001, there were signs of some movement in China's position. Former premier Zhu Rongji, for example, reportedly indicated to foreign and domestic scientific and policy elites in 1998 that China should find some constructive role to play on the issue of climate change. In addition, he requested that an international group of environmental and economic experts explore potential strategies for China to reduce its carbon dioxide emissions. Meanwhile, the growing role of the State Economic and Trade Commission (SETC) in the wake of the 1998 administrative reforms (see chapter 4) reportedly gave greater voice to those in the government who were interested in pursuing a more proactive climate change policy in order to gain access to new technologies from abroad. According to one Chinese scientist, there was increasing dissatisfaction with the recalcitrance of the Ministry of Foreign Affairs and the State Development and Planning Commission (SDPC). The sentiment was that China was missing significant opportunities to advance its technological know-how because of a reluctance to cooperate on climate change.

In 2002, China took a significant step forward at the UN World

Summit on Sustainable Development in Johannesburg when it announced that it had ratified the Kyoto Protocol to curb greenhouse gas emissions. As a developing country, China is eligible under the Clean Development Mechanism (CDM) to earn credits by undertaking emissions-reduction activities;[11] however, it is not required to meet any emissions targets or timetables. Nonetheless, China's taking this first step means that it may later be pressured to agree to emissions reductions commitments. (Note that, even without any formal commitment to reduce its emissions, China reportedly reduced its emissions of carbon dioxide during 1996–1999 by restructuring its economy, switching to cleaner energy sources such as natural gas, and improving energy efficiency.)[12]

The cases of global climate change and ozone depletion suggest both the importance and the limitations of the international community's influence. It is clear that the international scientific and policy-making communities had little influence in determining the outcomes of China's decision-making processes regarding these issues. In the case of global climate change, the science has remained incidental to political decision-making. In the case of ozone depletion, although the scientific community played a role in domestic deliberations, it was not until China's financial demands were met that the country assumed a more proactive orientation.

However, interaction between the international community and domestic actors and institutions on climate change continues and may yet produce more substantial changes in both process and outcome. Moreover, the relatively high degree of institutionalization of a new, broad-based environmental bureaucracy with direct and extensive international ties suggests the potential for real policy change in the future.

Setting a New Agenda for International Environmental Cooperation

Global climate change and ozone depletion brought to the fore a new global discourse on issues related to the environment and development. These issues were given further impetus by the 1992

UN Conference on Environment and Development (UNCED), which focused attention on the broad issue of how best to integrate environmental objectives with economic development goals. Inspired and impelled by these three sets of global environmental discussions, Chinese environmental and scientific elites along with their international counterparts launched a domestic and international offensive to advance the cause of environmental protection in China.

In the run-up to the Rio conference, Chinese environmental and scientific officials held a series of meetings to educate and pressure their planning and industry colleagues about the importance of incorporating environmental objectives into China's plans for economic development. With significant encouragement from Martin Lees, a former UN official, and funding from the international community, the Chinese leadership sponsored a three-day high-profile international meeting in October 1990, hosted by environmental, scientific, and social science communities, represented most prominently by then NEPA head Qu Geping, State Science and Technology Commission (SSTC) chairman Song Jian, and president of the Development Research Center of the State Council Ma Hong. The International Conference on the Integration of Economic Development and Environment in China brought together Chinese leaders and international experts to discuss issues pertaining to the program's theme, "China and the World in the Nineties." This was the first of several international conferences hosted by Beijing to focus on environmental concerns, and participants from outside China included representatives from the World Bank and UNDP, business executives from Shell and Sumitomo, and the heads of NGOs like the Rockefeller Foundation and the World Wide Fund for Nature.

Qu Geping used his position as conference chairman to bring international pressure to bear on the somewhat less environmentally inclined ministry representatives from the Ministry of Energy, the State Planning Commission, and the Ministry of Foreign Affairs. The officials from these agencies stressed either the economic imperatives of advancing development in China or the primary responsibility of the developed countries—which had consumed, they argued, global resources and degraded the environment in their drive to industrialize—to address pollution issues.[13] By contrast, representatives from the Ministry of Agriculture, the Ministry of Water Conservancy, and the Development Research Center of the State Council claimed that China's pollution stemmed primarily from China's inappropriate pricing system for natural resources, the traditional view that resources are inexhaustible, and poor local management techniques.[14] In his summation, Qu Geping delineated the industrial and environmental arenas in which China should take stronger action to improve environmental protection measures.[15]

Soon after this meeting, then chairman of the State Environmental Protection Commission Song Jian publicly set out an alternative formula to the one typically articulated by Chinese leaders. Rather than simply pushing forward on economic development and then playing cleanup afterward, Song argued for balancing the two simultaneously, or even slowing down economic growth in order to protect the environment: "As we develop the economy, we must guarantee a balanced ecological environment and maintain in good order our natural resources so that future generations will have their rightful heritage. To this end, we should be ready to pay more or, if necessary, slow down the economic development."[16]

This 1990 meeting also launched the China Council for International Cooperation on Environment and Development (CCICED), which many Chinese environmental actors consider one of the most prestigious and effective forums for international environmental cooperation. The first formal meeting of the CCICED, which was coordinated by China's NEPA and the Canadian International Development Agency, was held in April 1992. Chaired by Wen Jiabao until he became premier in 2003, CCICED has served as a forum for international and Chinese experts to exchange views and develop recommendations for Chinese leaders.[17]

The advent of the UNCED in 1992 gave a dramatic boost to the establishment of a formal institution for international and Chinese cooperation on environmental protection. Although the Chinese delegation to the UNCED, led by Premier Li Peng, not surprisingly reflected China's traditional values and rhetoric,[18] the conference had a profound effect on environmental policies, institutions, and thinking in China.[19] Within one year of the UNCED, China became the first country to develop its own national Agenda 21 based on the global Agenda 21 action plan to promote sustainable development. Here, too, the international community was instrumental in shaping the domestic policy process. The United Nations Development Programme proposed the idea, provided financial support, offered international expertise, reviewed the proposed priority projects, and arranged the international donor meetings.[20]

Chinese environmental actors called directly on the international community to assist in supporting Agenda 21. During the summer of 1994, SSTC vice president and daughter of Chinese leader Deng Xiaoping, Deng Nan, stated:

> The development of the economy and technology has made it possible for us to increase investment to deal with environmental problems, but economic development in China at present is basically resource oriented. It is possible that development will bring about destruction of the ecology and worsening of the environment. If environmental problems are ignored in the process of development, economic development will be severely hampered. We should extensively launch international cooperation.[21]

In July 1994, shortly after Deng Nan's statement, China held a second high-level meeting to garner international financial support for the national Agenda 21 projects. Representatives from twenty countries, international institutions, and businesses agreed to support approximately forty of the sixty-two high-priority projects that the Chinese had outlined, including projects focused on cleaner production, sustainable agriculture, pollution control, clean energy, and the development of communications. For Agenda 21 to succeed, the international community, according to Song Jian, would have to supply 30 to 40 percent of the estimated $4 billion cost.[22]

China's participation in the UNCED was also a catalyst for institutional change, contributing to the development of environmental NGOs and the formation of an Administrative Center for China's Agenda 21, directed initially by the SSTC and the SPC, but later directed primarily by the SDPC.

Perhaps most important, for the Chinese leadership, the UNCED also served to reinforce and, for the general public, introduce the concept of *sustainable development*. The term now provides an accepted framework within which Chinese officials press for the incorporation of environmentally sound practices into future economic development.[23] Throughout many regions of China, local governments have developed their own Agenda 21 bureaus to ensure that sustainable development principles are incorporated into urban planning.[24]

The international community has responded to the rise of Chinese environmentalists with both institutional and financial support. In many cases, different international actors send the same message. For example, CCICED has advocated enhancing the State Environmental Protection Administration's (SEPA) authority. At the same time, the World Bank's 2001 report, *China: Air, Land, and Water*, explicitly calls not only for raising SEPA's profile but also for reconstituting the State Environmental Protection Commission, the environmental oversight body that was dismantled in the 1998 administrative reforms.[25]

China's efforts to attract international investment for environmental protection have been extraordinarily successful. According to one estimate, fully 80 percent of China's environmental protection budget is derived from abroad.[26] Overall, China is the largest recipient of environmental aid from the World Bank and the Global Environmental Facility and has received extensive assistance from the ADB and UNDP.

The World Bank during 2000–2001, for example, provided more than $1 billion in loans for projects on such environmental problems as water conservation, improving the environment in Beijing and Chongqing, and controlling pollution in the Huai River Valley. New lending in 2002, however, totaled only about $110 million, although two significant projects, including a $200 million urban environment project for Shanghai, were delayed. For 2003, only two environment-related projects were funded, including the delayed Shanghai urban environment project. The $350 million allocation for these projects, moreover, paled in comparison to the $650 million approved for highway projects. On a smaller scale, during the past decade, the ADB has set aside more than $4 billion for environment-related projects in China; in 2002, it provided more than $200 million in loans and grants for flood management, wastewater treatment, and support for a China Environment Fund.

Japan is the largest bilateral provider of environmental assistance to China and has made such assistance the number one priority within its overseas development aid (ODA) budget. Japan has a strong interest in shaping China's environmental practices; it is affected by both the acid rain and dust that emanates from China and by such issues as fishing rights. Japan also perceives a direct economic benefit from its assistance in terms of opportunities for environmental technology transfer. In 1999 Japan's assistance totaled $1 billion, and over two-thirds of its projects were devoted to environmental protection. In addition, during 1995–1999, Japan sent five hundred experts under environment/technical cooperation efforts to China.[27]

The United States provides extensive policy advice and technical training opportunities to China, primarily through NGOs. (Its formal bilateral assistance is sharply constrained by political considerations.) The newfound influence of the Chinese and international environmental communities has also been evidenced by the inclusion of environmental issues—both global and domestic—on the agenda of leaders' summits in 1998 between President Jiang Zemin and President Clinton and between Jiang and Japanese Prime Minister Obuchi.

Despite these successes, China's interaction with international lending agencies and its bilateral relationships are plagued with political problems. The World Bank, for example, has sharply reduced funding to China. In part, this is because China no longer qualifies for International Development Assistance (IDA) funds, which are provided only to the poorest countries. The other difficulty, however, has been the strong control exerted by the SDPC over which projects will receive World Bank funding, which creates great difficulty for the Bank in establishing new partnerships with Chinese entities. Over the years, the same Chinese cities have repeatedly received Bank money for a wide array of projects. According to one Bank official, there is a rationale for this. Shanghai, for example, has strong institutional capabilities, well-trained people, and lower costs of doing business than in many other regions of the country. Most important from the perspective of the SDPC, however, has been that Shanghai can repay its loans. However, in the eyes of this Bank official, many vital projects in less-developed regions not only go unfunded because of the lack of local money but also cannot attract World Bank money. There has been little opportunity to overcome the SDPC's power because it has effectively thwarted direct local collaboration with Bank officials.

Even powerful, environmentally proactive actors such as the Ministry of Finance or SETC—which until its dissolution in the 2003 administrative reforms (which, like the 1998 reforms, resulted in the merger, dissolution, and addition of various ministries and commissions) had broad oversight for energy conservation, efficiency, and renewable energy such as solar and wind—had to negotiate with the SDPC if the programs involved World Bank funding through the International Bank for Reconstruction and Development money. Only if the program dealt with climate change, thus guaranteeing grant money from the Global Environmental Facility, was the SDPC not involved in the deliberations. Similarly, the UNDP increasingly sought to fund projects through the Global Environmental Facility in order to avoid working through its Chinese counterpart agency, the more development-oriented Ministry of Foreign Trade and Economic Cooperation (MOFTEC).[28] With the reconstitution of the Chinese bureaucracy in 2003, it is possible that the State Development and Reform Commission (SDRC) (the former SDPC merged with part of the Structural Reform Office and part of the SETC) along with the new Ministry of Commerce (a merger of part of the SETC with MOFTEC) and their new heads, Ma Kai and Lu Fuyuan, respectively, will be more environmentally proactive than their predecessors at the SDPC and MOFTEC. There is little in their past histories to suggest either experience with or interest in the environment, however, as with Qu Geping, such interest may emerge.

China's relationship with Japan presents its own set of issues. The Japanese government has decided that, given the overall strength of the Chinese economy, Japan's ODA will be targeted almost exclusively at environmental projects. It will no longer support large-scale infrastructure development. According to some Japanese analysts, however, the Japanese people are concerned that Japanese aid frees China to devote its own funds to developing its military strength or space program, and they question the value to Japan of such assistance. Moreover, the prolonged economic downturn in Japan prompted the Japanese public in 2002 to express growing impatience with all forms of overseas development assistance.

Unlike in the case of Japan, extensive U.S. environmental diplomacy with China has not been matched by an equally impressive set of collaborative projects. The U.S. Department of Energy (DOE) has had difficulty exploiting the potentially vast China market for U.S. cleaner coal technologies. After DOE requested $1.4 billion for a global cleaner coal technology program, approximately one-quarter of which was for China, the U.S. Congress refused to fund the program, calling on the Overseas Private Investment Corporation (OPIC), the Export Import Bank (EX-IM Bank), and the U.S. Agency for International Development (USAID) to fund it instead. However, OPIC and USAID are barred for political reasons from doing business in China. The result has been a disappointingly low level of cooperation, limited largely to discussions and U.S. Environmental Protection Agency (USEPA)–funded DOE projects. The lack of assistance, and especially of financing mechanisms for U.S. corporations, also means that U.S. firms cannot compete with their European and Japanese counterparts.

Finally, the international community may also play a role in spurring China's leaders to take action on environmental issues because of concern over outsiders' perceptions. For example, when former president Clinton toured Shanghai with then mayor Xu Kuangdi in 1998, Mayor Xu was greatly embarrassed by the amount of garbage they encountered. Shanghai soon began plans to organize a massive citywide collection and recycling program. Similarly, Beijing's efforts to clean up local air quality were sparked not only by the local leaders' interest in the country's October 1999 Fiftieth Anniversary celebration but also by the visit of the International Olympic Committee screening committee in spring 2001. As a result, Beijing implemented some technologies with long-term implications. For example, the government replaced 21,000 coal-burning heating units with cleaner-burning gas-fired ones in factories, municipal offices, and restaurants.[29]

Beyond the direct aid and assistance provided by the international governmental community, it is China's integration into the world economy that may provide the greatest opportunities, as well as the greatest challenges, for accomplishing a long-term revolution in environmental protection practices in China.

Economic Reform: Challenges and Opportunities for Environmental Cooperation

China's current stage of industrialization, transition to a market economy, and integration into the international economy has contributed to the pollution and degradation of the environment over the past three decades. Yet it also has provided a substantial opportunity for rethinking and reorienting the country's approach to environmental protection.

The same attributes needed for a successful market economy and integration into the global economy, including transparency, rule of law, managerial expertise, and a premium on efficient production, are the keys to realizing the environmental goals outlined in China's Agenda 21. As China's leaders press forward with economic reform and integrating China into the global economy, the international community has taken the opportunity to advance environmental cooperation in four areas: policy design, capacity building, technology transfer, and enforcement and incentives.

Policy Design

The international community and China have developed an extensive range of cooperative activities focused on policy reform. Many of these efforts target China's energy sector, including plans to develop new strategies and laws at the national level, as well as local initiatives, such as developing energy efficiency building codes in Chongqing.

Foreign governments, international governmental organizations (IGOs), international nongovernmental organizations (INGOs), and multinationals have all established partnerships with Chinese ministries and other institutions focused on these issues. The World Bank and the State Development and Reform Commission are collaborating on a cleaner coal strategy; British Petroleum (BP) is supporting research at Qinghua University to develop an energy strategy, as well as a ten-year $10 million research program under the Chinese Academy of Sciences to consider strategic and technical issues relating to the development of cleaner sources of energy including natural gas and the introduction of hydrogen as a fuel; and the USEPA, Pacific Northwest National Laboratory, SDRC, and the University of Petroleum in Beijing are collaborating to explore the potential for natural gas utilization in China. Already, a gradual increase in the use of natural gas in Beijing has contributed to an improvement in air quality. The Natural Resources Defense Council has begun several long-term projects to develop building codes for the Yangtze River Basin region, a painstaking effort that involves the preparation of detailed manuals; training for a range of stakeholders, including enforcement officials, designers, builders, and component manufacturers; the development of financial incentives; and improvement of China's domestic capability to manufacture energy-efficient building materials.[30]

Major new infrastructure projects, like the Three Gorges Dam and the West-East pipeline (from Xinjiang to Shanghai), if effective, also promise to help reduce China's long-term reliance on coal. China took an important step in 2001 by banning lead from its gasoline, the result of a concerted, more than six-year lobbying effort by international actors such as the World Bank and General Motors, coupled with support from domestic Chinese actors such as SEPA, the Ministry of Science and Technology, local health officials, and eventually, the major oil bureaucracy, China Petroleum and Chemical Corporation (SINOPEC).

China's transition to a market economy has also opened the door to a wide range of new market-based policy approaches. At a CCICED meeting in mid-October, 2001, in Beijing, Asian Development Bank vice president Joseph Eichenberger reinforced the message that measures such as tradable permits, carefully targeted use of the tax system, and appropriate pricing can be more effective than traditional regulatory approaches.[31] The CCICED itself claims as part of its impressive list of accomplishments convincing China's policy makers of the importance of price reform for energy and water, encouraging the advancement of cleaner production, and supporting incentives for renewable energy use.[32] Chinese environmental protection officials have stated that they will increasingly rely on financial incentives and the market rather than simply fines[33] and the court system to improve the environmental situation. In the water-scarce northern city of Zhangjiakou, Hebei Province, for example, officials, acting on the advice of the ADB, raised the price of water 40 percent, but varied the price according to the type of consumer. In one year, water consumption in the city fell by nearly 14 percent, primarily due to reductions by factories that had finally instituted water recycling.[34]

Already, some international actors such as the U.S.–based NGO Environmental Defense have developed pilot projects employing market-based tools. To assist in helping China meet its target of reducing sulfur dioxide emissions by 20 percent, for example, Environmental Defense initiated a project in 1997 to institute a system of tradable permits for sulfur dioxide. This system, which Environmental Defense introduced in the United States in 1990, establishes an overall level of permissible emissions for a given region, assigns targets for individual companies, and then permits the enterprises to buy and sell the rights to pollute. Firms whose emissions fall below the permitted level can store the excess quota for future use or sell it to others who cannot meet their emissions levels. Not surprisingly, the Chinese have encountered some hurdles in their efforts to utilize a market-based tool in a transitional economy. As William Alford notes, the hybrid nature of the Chinese economy often stymies policies that otherwise might be important mechanisms for environmental protection:

Adherence to the "polluter pays" principle and the establishment of a workable system of tradable discharge permits presumes more in the way of market mechanisms overseen by independent regulatory authorities than is now available in China or likely to be in the foreseeable future. . . . The "polluter pays" principle, for example, depends on such variables as readily identifiable corporate entities that operate under meaningful budget constraints, the free flow of information needed to make a market, and a clear dividing line between those who are regulated and those doing the regulating.[35]

Researchers involved in the experiments for tradable permits for sulfur dioxide have also noted that Chinese officials and enterprise directors have only a weak understanding of the concept of property rights. There also is no established mechanism for property rights transfer in China and thus no logical transaction agent (i.e., a person to hold the money in escrow during the transaction). Still, with a labor-intensive education campaign, flexibility, persistence, and willingness to modify some aspects of the Western experience to fit China's needs, the project has expanded from one test case to four provinces, three cities, and one electric generating company that account for roughly one-third of China's sulfur dioxide emissions.

Capacity Building

The case of the tradable permits for sulfur dioxide underscores the importance of capacity building within China if environmental cooperation is to be effective. Without the proper education and training for Chinese partners, and the appropriate coordination among them, many policy initiatives will never be fully implemented.[36] According to an assessment of ADB projects in China, "More than half of the pollution discharges can be reduced at many enterprises through strengthening management rather than upgrading equipment."[37] Capacity building, therefore, has become a central element of most international cooperative ventures.[38]

BP, for example, conducts extensive management training sessions and exchanges on environmental issues. After buying part of Sinopec, China's second-largest oil producer, BP moved quickly to improve Sinopec's environmental practices. Sinopec officials stated that they would spend $146 million in 2001 to improve their environmental and safety practices. According to the head of Sinopec's Safety and Environment Bureau, the underlying logic for doing this is simple: introducing a health, safety, and environmental management system is "good for our corporate image and creates a good foundation for entering world markets."[39] Other companies, such as General Electric and General Motors, also hold wide-ranging training programs for their Chinese plant managers, covering legal issues, environment, health and safety questions, audit skills, and work plans for specific issue areas such as maintaining air and water quality, pollution prevention, and industrial hygiene.

Given the long-term investment in capacity building and training necessary for many partnerships, multinationals and other international actors often stress the importance of selecting the right partner for a cooperative venture.[40] General Motors, for example, searched five years before selecting Shanghai Automotive as its Chinese partner.

Still, officials from one U.K.–based multinational believe that when a Chinese company or ministry is looking for cost reduction in a planned venture, the first thing to be eliminated from the budget is training. Moreover, they have found that although there has been a "sea change in attitudes" within Chinese enterprises over the past decade, these new understandings are "not always reflected in implementation." While the company's Chinese partners, for example, have been well-briefed in the International Standards Organization (ISO) 14001 standard, by which an organization can be certified as having attained a certain international standard for its environmental management system, they do not always implement the guidelines fully.[41]

Bridging the gap between policy design and implementation in China also requires extensive bureaucratic coordination, a challenge often not met adequately. Both Chinese and international participants of CCICED, for example, believe their environmental recommendations are often not adequately implemented because there is little to no participation by the ministries responsible. As one Chinese CCICED participant has noted, "When it comes to the relevant ministry, due to lack of institutional relationship between the

Council and this Ministry, it's likely that neither the minister nor its key departments or divisions know anything whatsoever about the CCICED." Similarly, the ADB has found that weaknesses in interagency coordination "frequently lead to turf wars among agencies."[42] Instituting cleaner production,[43] for example, has required coordination among SDPC (responsible for policy and budget), SETC (responsible for industries), the National People's Congress (NPC; responsible for legislation), the Ministry of Science and Technology (MOST; responsible for technology transfer), SEPA (responsible for environmental management), and the Ministry of Agriculture (responsible for TVE [township and village enterprises] management).[44] Getting all these organizations to talk to one another and coordinate their efforts is difficult, although this may be simplified by the 2003 bureaucratic reforms.

Technology Transfer

China's economic development and reform also introduce the potential for significant environmental advances through the introduction of new technologies. The 2001 World Bank/SEPA assessment of the state of environmental affairs in China notes, "The switch to a more competitive, demand-driven industrial sector is resulting in increased earnings retention and re-investment. This is increasing technological innovation and resource use efficiency, allowing more industrial growth to be achieved at less environmental cost."[45] Moreover, local officials often express the belief that technology is the answer to their environmental problems.

To facilitate this process, China instituted a cleaner production program in 1993 and ISO 14000 environmental management system certification procedures in 1997.[46] Both international governmental organizations and multinationals have developed a number of impressive cooperative ventures employing cleaner production. Successful ventures, not surprisingly, combine policy change, technological innovation, and capacity building. The ADB, for example, has implemented two large-scale cleaner production projects, one at Hefei Chemical Works and the other at Hefei Iron and Steel Company in Anhui Province. Both projects have contributed to increased productivity and efficiency and lower levels of pollution.[47] General Motors has instituted cleaner production in its joint venture with Shanghai Automotive and has committed to building and importing vehicles with advanced emission technologies such as fuel injection and catalytic converters. To ensure the viability of these technologies, GM instituted a training program for service centers on how to properly service and maintain these technologies; launched a cross-industry campaign urging the production and use of clean fuel in China; and organized workshops for the Chinese government, oil industry, auto parts industry, other businesses, and universities to discuss clean fuel for advanced technology.[48]

Establishing new institutions to facilitate technology transfer and environmental education is also an important means of enhancing China's environmental protection efforts. The World Bank, for example, has supported the development of energy service companies (ESCOs)—independent companies that advise Chinese companies on buying, installing, and maintaining energy-efficient technologies. ESCOs are then paid with the financial savings that the technologies brought to the companies. There were more than a dozen ESCOs in operation by late 2002.[49]

Many multinationals, however, have been stymied in their efforts to transfer and/or employ environmental technologies successfully. Even when coupled with capacity building and policy reform, the effectiveness of new technological advances is impaired by China's weak enforcement capacity and incentive system. Case after case demonstrates that "technology is not a substitute for enforcement."[50] Many firms possess the technological capacity to address their pollution problems, whether it is waste water treatment or electrostatic precipitators to control sulfur dioxide emissions, yet avoid using this equipment because they believe it is too costly.[51]

Enforcement and Incentives

China's feeble capacity to enforce its environmental laws and regulations is well known both within and outside the country. The

2001 World Bank/SEPA report notes, "Cleaner Production and ISO 14000 programs are not a substitute for the budgets and manpower needed to effectively enforce pollution laws and regulations."[52] A 2000 SEPA study, too, noted this enforcement vacuum: "[O]ne-third of the pollution abatement facilities installed in Chinese enterprises are operated 'inefficiently' (i.e., they are only switched on during inspections), one third are never operated because operators are worried about associated operating costs, and a further third are operated as required by regulatory standards."[53]

For the international community, enforcement challenges take several forms. For example, officials from the British company, National Power, have argued that China holds different types of power plants to different standards.[54] Many ventures have stalled or dissolved over the failure of China to protect intellectual property.[55] And other firms have found, as the 2000 SEPA study noted, that their Chinese partners elect not to utilize some end-of-the-pipe technologies because of their perceived additional cost. Even if multinationals provide the necessary pollution control equipment, Chinese firms may not fit it at industrial sites because the cost is much higher than the small financial penalties they will have to pay for exceeding pollution limits.[56] BP officials, too, have commented that Chinese enterprise directors remain concerned that international best practices add costs and create delays.

Not surprisingly, weak enforcement may affect the willingness of firms to transfer their best technology. Multinationals are understandably reluctant to employ their advanced technology, which typically adds more to the start-up costs of the venture, if they don't believe that inferior technology will be penalized.[57] As some foreign companies have argued, there would be a far greater incentive to transfer cleaner coal technologies to China if regulations favored these technologies over dirtier alternatives. As matters stand, however, multinationals believe that any cleaner power plants they own or build will be undercut by cheaper, dirtier plants owned by provincial power companies.

The challenge of enforcement is matched by the need for development of a stronger incentive system for enterprises to invest in technical measures to improve plant efficiency.[58] As the World

Bank/SEPA report suggests, for cleaner production to be adopted, enterprises need incentives such as resource shortages, higher prices, and increased regulatory pressure. To date, cleaner production has been limited to regions with strong incentives, such as the water-scarce areas of northern China,[59] or to joint ventures in which the technology was provided by the international partner.

Even if a Chinese enterprise employs the most advanced environmental technologies, the local infrastructure as well as the capacity of the enterprise management may be insufficient to support their use. As one analysis reports, "The latest environmental technologies are often inapplicable in China, where existing systems are often many years out of date. Grafting new technologies onto China's old systems may lead not only to sub-optimal output but even to outright failure."[60] Dupont and Dow officials, for example, found that "unstable power supply often disrupted their ultra sensitive high-tech chemical manufacturing process. Computerized production components failed, and onsite technicians were often ill-prepared to handle the consequences."[61] In one case, the company was compelled to bring in experts from abroad, which increased the cost of the venture far beyond initial expectations. Company officials were especially unhappy because the Chinese government had insisted that only the most advanced technology be used, and the venture had been predicated on the government's assurances of appropriate infrastructure support.[62]

China's continued evolution to a full market economy, well integrated into the global economic system, provides a number of opportunities and challenges for environmental cooperation. However, given the range of potential institutional, policy, and technological impediments to effective collaboration, an integrated approach, which supports a variety of assistance needs, is needed to achieve the best results.[63] This means that all four potential areas of cooperation—policy design, capacity building, technology transfer, and enforcement/incentives—must be targeted for international involvement. As China's economy becomes further integrated into the global economy, there is also the potential for environmental objectives to be advanced by environmental activities or requirements embodied in the various trade regimes to which China accedes.

Reinforcement by Trade Regimes

The tensions inherent in integrating environmental protection with international trade are evident even in the most environmentally oriented international declarations. For example, Principle 12 of the Rio Declaration states:

> Trade policy measures for environmental purposes should not constitute a means of arbitrary or unjustifiable discrimination or disguised restriction on international trade. Unilateral actions to deal with environmental challenges outside the jurisdiction of the importing country should be avoided. Environmental measures addressing transboundary or global environmental problems should, as far as possible, be based on international consensus.[64]

Some sectors in China have already been affected by green requirements in international trade, including textiles, agricultural products, and wood crates for packing that have been banned by the United States and Canada.[65]

Yet regional and global trade regimes such as APEC and WTO have the potential to reinforce positive trends in China's development and environment trajectory. China's accession to the World Trade Organization in December 2001 prompted the government to establish an interministerial working group under the auspices of the former Ministry of Foreign Trade and Economic Cooperation to think through the environmental implications of China's WTO accession.

WTO accession will have real and significant impacts for China's future economic development and thus the environment. For example, it is likely that China will shift production away from areas of comparative disadvantage such as grain production and intensive forestry production with positive environmental effects. (China's growing timber imports, as noted in chapter 3, however, are exerting a negative effect on the environments of other countries such as

Burma.) Increased imports of paper and pulp products might well lead to the closure of China's own highly polluting paper and pulp industry. In addition, a high level of inefficiency in some of China's manufacturing industries such as autos, chemicals, and machinery should engender greater opportunities for imported products as well as multinationals producing in China.[66] Other potential positive changes resulting from WTO accession include publication of environmental laws and regulations that may affect trade, international insistence on greater legal transparency and effective enforcement mechanisms in China,[67] and the continued closure of inefficient state-owned industries.

The WTO has already raised the environmental stakes for some manufacturers in China. Environmental advocates in China are discussing the necessity of strengthening the country's sanitary and phytosanitary regulatory regime to ensure that Chinese exports meet others' standards and that China will provide adequate protection for its own people and environment.[68] In February 2002, the European Union (EU) banned Chinese poultry, shrimp, prawns, and several other food products on the grounds that the food was contaminated with a prohibited antibiotic.[69] Previously, the EU had barred Chinese-manufactured tea from some regions based on the pesticides used. Actions like these will inevitably push Chinese industry toward compliance with worldwide standards.

On the other hand, China's accession to the WTO will strengthen the export potential for some sectors such as textiles, toys, and food products (e.g., livestock, fruits, and vegetables).[70] Some exports are projected to grow significantly; textiles, for example, are expected to increase by almost 64 percent during 2000–2005. This will likely have detrimental effects on water quality. Especially worrisome is the potential for the Chinese government to encourage the expansion of environmentally harmful industries such as tin mining and production.[71] Similarly, as noted earlier, car ownership is likely to increase between 5 and 10 percent annually, leading urban air quality to decline further.[72]

While APEC, unlike the WTO, does not impose requirements or binding commitments on its participants, it does provide a "stocktaking exercise" in which each member economy reports on steps it has taken to promote sustainable development in several key areas, including cleaner production, protection of the marine environment, food, and energy.[73] In addition, the APEC energy ministers devote substantial attention to the clean use of energy, renewables, energy efficiency, and so on, even encouraging member economies' participation in activities to reduce greenhouse gases.[74] Finally, APEC contributes to regional environmental workshops. In 2000, for example, China and the United States sponsored an APEC workshop on the disposal of aging or damaged equipment used during offshore oil and gas exploration to prevent marine pollution. It was attended by 150 government, private sector, and university participants from throughout the Asia Pacific.[75]

Still, the environmental component of APEC appears to be losing rather than gaining steam over time. The number of projects related to sustainable development declined from sixty in 1998 to forty-two in 1999.[76] The environment ministers met last in 1997, despite pledging to meet annually, and Chinese foreign minister Tang Jiaxuan's lengthy post-APEC summit interview with Xinhua News Agency in October 2001, made only glancing reference to the summit's promotion of a "sustainable development of the Asia-Pacific regional economy,"[77] which well may not even have been referring to the environment.

Instead, regional environmental issues are usually addressed through small commissions or organizations involving only those countries directly affected. South Korea and Japan, for example, work directly with China to monitor acid rain. The same three countries also continuously negotiate marine fisheries agreements. More difficult for regional cooperation are issues that involve sensitive sovereignty issues, such as control over resources in the South China Sea or utilization of the water resources of the Mekong River. In January 2002, China began construction of its third hydroelectric power dam on the Mekong; five more are planned by 2020. China's actions have caused great consternation among the affected countries downriver, including Burma, Laos, Thailand, Cambodia, and Vietnam; Vietnam has accused China's dam construction of cutting the flow of water, resulting in saltwater intrusion into the Mekong Delta, which is home to about half of Vietnam's agricultural output.[78] While China did begin to participate in the Greater Mekong Subregion Commission that is focused on development issues, despite significant international pressure, China has refused to join the Mekong River Commission that coordinates environmental issues among these states. In 2003, in a notable concession to its downstream neighbors, China began providing information on the river flow and water levels of the Mekong.

Thus, China's growing participation in international trade regimes, its development of domestic environment and development institutions to further cooperation with the international environmental community, and its transition to a market economy all contribute to an increasingly dense network of linkages on issues related to environmental protection. There are few if any potential risks to China in such cooperation. But, as China seeks to invite foreign participation in the full range of its environment and development activities, its continued penchant for large-scale development campaigns has opened the country to mounting criticism from both within and outside China.

The Great Campaigns: Reconciling Environment and Development

Three Gorges Dam

The construction of the Three Gorges Dam has received perhaps the lion's share of international (as well as domestic) criticism. Sun Yat Sen initially proposed the project back in 1919, but the process of realizing his vision began in earnest only in the 1980s and early 1990s.[79] The dam is a massive undertaking that will provide China with 18,000 megawatts of energy, more than 10 percent of China's total electricity needs. It will stand 600 feet tall and create a reservoir more than 360 miles long and 175 meters deep. In a two-stage process, the dam will first provide electricity to a region in the immediate vicinity of the dam; by 2010, however, the government plans to transmit power as far north as Beijing and as far south as Hong Kong. Whether the demand for energy will be there at that time has yet to be fully addressed.

The Chinese scientific community has been divided on the merits of the dam. Although the U.S. government provided technical assistance for the Three Gorges Dam during 1984–1993, many international experts have since condemned the project and serious debate occurred within the National People's Congress during the early 1990s. At the time, those opposed to the dam, both within and outside China, argued that the project was unsound on several grounds: (1) it would force the relocation of over one million Chinese citizens; (2) it would flood precious arable land and destroy ancient cultural relics and historic cities such as Fengdu, also known as the "ghost" city; and (3) the reservoir created by the dam would become filled with sediment and prevent the passage of some ships. Nonetheless, the Chinese government elected to proceed with the dam and began construction in 1994.

At first, international financial support for the dam was scanty. The World Bank refused to help,[80] and Chinese plans to issue bonds on the international financial market were dashed by the lack of interest. However, the Three Gorges Project Development Corporation began to raise money through the China International Capital Corporation, which is 35 percent owned by Morgan Stanley. Moreover, U.S. investment firms, including Morgan Stanley, have helped underwrite $830 million in China Development Bank bonds, whose top loan commitment is the Three Gorges Dam.

Meanwhile, international forces began to fight the dam. International Rivers Network, the international nongovernmental organization, undertook a broad-based campaign to force Morgan Stanley to withdraw from providing any assistance in fund-raising for the dam, including boycotting the company's Discover Card. In addition, Trillium Asset Management, a Boston-based firm that promotes "socially responsible" investing, has been active since 1995 in exerting pressure on several other financial institutions, including Citigroup and Chase Manhattan Bank, not to contribute assistance in any form to the dam, and has met with some limited success, as when Bank of America agreed not to commit any direct lending and to "carefully weigh any transactions that might indirectly benefit the dam."[81] In addition, in part as a result of Trillium's efforts, Citigroup has begun to incorporate environmental, human rights, and public relations risk criteria into its lending and under-writing deals.

Contributions to the capital construction of the dam also ran into some roadblocks. While many governments helped their domestic firms win bids on various aspects of the project by providing finan-cial assistance and insurance through their export credit agencies, the United States, led by the U.S. State Department, withheld sup-port through its Export-Import Bank.

As construction of the Dam has progressed, new problems have arisen, giving rise to even greater international and domestic con-cern. The project has been plagued by corruption. According to one report, local officials involved in managing the construction and re-settlement process have embezzled as much as $60 million. Shoddy construction that caused a bridge to collapse and kill several people led former premier Zhu Rongji to term the construction "tofu" en-gineering.[82] He then invited foreign experts and engineers to oversee the construction process to ensure that top-quality practices were enforced.

Social unrest surrounds the resettlement process. Since the late 1990s, there have been many demonstrations involving hundreds of farmers who believed they were being inadequately compensated during the resettlement process. Probe International and Human Rights Watch have joined International Rivers Network in moni-toring the resettlement process and the local political situation around the dam and have issued several scathing reports regarding the corruption that has plagued the resettlement efforts. In July 2002, when fifty-six migrants from Gaoyang, located 225 kilome-ters from the dam site, attempted to travel to Beijing to lodge com-plaints concerning the lack of compensation for their resettlement, they were accused of being Falun Gong supporters and arrested at the direction of the local county leadership. They were released after their plight was brought to the attention of officials from the Three Gorges Project Construction Committee.[83] Many older Chi-nese, too, are reluctant to leave their homes, many of which have been in their families for centuries. Indeed, all along the Yangtze, beautiful stone houses that seem to be embedded in the rocky land-scape stand shuttered as new, often spanking white, multistory apartment buildings sprout at higher elevations in newly developed cities. The deserted towns have also spawned another industry: private "grave robbers" search in the villages for ancient stone tablets that might provide historical records. While about one-third of the resettlement has been accomplished, almost 1.3 million people still await relocation. Chongqing, alone, must still resettle almost 500,000 people and more than 500 enterprises before 2009. In spring 2003, Gan Yuping, former deputy mayor of Chongqing and currently vice-director of the Three Gorges Project Construction Committee, stated that there were "enormous" problems surrounding the resettlement effort, including a lack of adequate farmland, funds, and job creation programs.[84]

Yet another concern is the pollution generated by the towns and cities surrounding the reservoir. Less than one-third of the industrial and one-tenth of the urban domestic wastewater is treated before being dumped into the area that the reservoir will occupy,[85] and the official government English-language newspaper, *China Daily*, reported that six million tons of rubbish and ten million tons of solid industrial waste are dumped into the Three Gorges area and the upper reaches of the Yangtze every year; more than twenty million tons of garbage are dumped in the areas just around the reservoir. Sailing down the Yangtze during summer 2002, I was not surprised to see mounds of garbage floating all along the river from Chongqing to Wuhan. Of even greater concern to many in China, however, is that the government will not have enough time to clean up all the factories and coal mines along the river banks before the reservoir is formed and that massive underwater pollution will result.

The Chinese government is well aware of these challenges, although it has tried desperately to downplay their significance. In April 2000, Zhang Guangduo, a senior Chinese scientist and principal examiner for the Three Gorges feasibility study done in the 1980s, wrote to Three Gorges Project Construction Committee Director Guo Shuyan to share his fears that the reservoir would become an environmental disaster. He argued that managing the impending environmental crisis would necessitate an investment of $37 billion.[86] The government has promised to earmark funds to clean up the water in the reservoir, and some water treatment facilities are now operating. SEPA director Xie Zhenhua announced in March 2003 that water pollution along the waterways in the dam region was higher than expected and that many industrial wastewater treatment projects were not meeting their targets.[87] However, shortly after the damming of the river in June 2003, Xie said he expects 100 percent of the factories in the cities around the dam to have water treatment facilities by the end of 2003.[88]

While the international NGOs have generated a significant amount of negative publicity for the dam and have had an impact on the thinking of the World Bank, the United States, and some investment firms, they do not appear to have exerted much influence, even indirectly, on Chinese practices.

West-East Pipeline

The $18 billion West-East gas pipeline has generated similar controversy. When completed, the 4,200-kilometer pipeline is scheduled to run from the Tarim Basin in Xinjiang to Shanghai. BP originally planned to bid on the construction contract, but it withdrew from the bidding in September 2001, citing business considerations. But environmental and social concerns may have had an impact on BP's decision. A range of NGOs, such as Tibet Vigil, Tibet Society, and the Free Tibet Campaign had been joined by the Tibetan government-in-exile in urging BP to withdraw from the pipeline because it represented a "significant escalation of China's exploitation of oil and gas on the Tibetan plateau and will accelerate China's policy of transferring Chinese settlers into Tibetan areas."[89] NGOs continue to call on BP to divest from its 2 percent stake in Petro-China, which is developing the pipeline. Moreover, at least one other multinational oil company that has entered the fray to compete for the right to develop the pipeline has become concerned not only about pressure from international NGOs but also about possible attacks on the pipeline by local ethnic minorities, who view the pipeline as another effort by the central government to exploit their natural resources for Beijing's own benefit.

Opening the West: Exploitation or "Ecological Construction"?

The West-East pipeline is only one part of a much broader effort of the Chinese government to develop the resources of western China. China's "Great Opening of the West" (*Xibu Da Kaifa*) campaign has perhaps more potential to draw fire from the international community (as well as from domestic quarters) than any other current campaign. Indeed, China's plans to open the west—one of the country's most important domestic initiatives at the outset of the twenty-first century—have raised a red flag throughout the country and the international community, highlighting the extent to which the Chinese government has opened itself to domestic and international pressures that may, in the end, contribute to reconfigure significantly the nature of the campaign.

Launched in 1999, the "Go West" campaign was designed to "reduce regional disparities and to eventually materialize common prosperity"[90] by developing six provinces (Shaanxi, Gansu, Qinghai, Sichuan, Yunnan, and Guizhou), five autonomous regions (Ningxia, Tibet, Inner Mongolia, Guangxi, and Xinjiang), and one municipality (Chongqing) in western China.

The area encompasses approximately 5.4 million square kilometers and a population of 285 million people (56 percent of the land and 23 percent of the country's population).[91] The region is rich in natural resources, including gold, oil, natural gas, and coal, but much of it is difficult to access. Moreover, unlike the coastal provinces, the region remains poorly connected to the outside world, and the regional infrastructure is weak. It receives only about 3 percent of the foreign direct investment that flows to China annually. Approximately 90 percent of the eighty million Chinese estimated to live in poverty reside in the west.[92] Guizhou, one of the poorest provinces in China, boasts an average per capita income of not quite 8 percent that of Shanghai.[93]

To develop the infrastructure of the west, Beijing has aggressively encouraged investment from overseas business and aid agencies, China's wealthy coastal provinces, and Hong Kong. Stated infrastructure needs include telecommunications, railways, airports, electric power grids, and the $15 billion West-East gas pipeline from

Xinjiang to Shanghai that will transport 12–20 billion cubic meters of natural gas annually beginning in 2007.

While the World Bank and ADB have each already targeted at least $1 billion to assist China in its efforts, the rest of the international community remains skeptical.[94] The head of the British Chamber of Commerce stated frankly, "Specifics are needed as the hinterland is a big place."[95] Even seventy-one CEOs from Hong Kong, who spent ten days traveling to Xian, Beijing, and Chengdu during May 2001, were hesitant to commit more than $30 million to the endeavor. Some of the CEOs even skipped a planned stop in Urumqi due to security concerns. An editorial in one Hong Kong newspaper noted the "poor fit" between the expertise of the Hong Kong businesspeople and the development needs of western China.

Some Chinese scholars doubt the practicality of the campaign as currently configured. For example, Chinese economist Hu Angang articulated several factors that he believes will limit the campaign's efficacy:

> First, the policies of the Ninth Five-year plan cannot effectively curb the widening of regional disparities; second, global economic restructuring weakens the comparative advantages of the western regions in agriculture, energy, [and] raw materials; third, the formation of a pattern in which supply exceeds demand [within] the domestic market has weakened the west's relative advantages in resource exploitation; and fourth, after China joins the World Trade Organization, the opportunities will outweigh the challenges in the eastern regions, but the opposite will apply in the west.[96]

Hu suggests that human rather than physical capital needs to be restructured first, by expanding job opportunities, relieving poverty, lowering the birth rate, and improving population quality (e.g., through education).[97]

The Chinese leadership also faces a critical test of its commitment to environmental protection. The region has already suffered severe ecological degradation from rampant deforestation, mining, overgrazing, and intensive plowing of cropland. And the Go West campaign will only exacerbate the situation unless environmental protection is made a priority.

In their speeches and reports, the top leaders have been careful to cite "ecological construction" as one of the five major tenets of the Go West campaign. As SEPA director Xie Zhenhua has stated, "In the process of developing the west, environmental protection programmes must be considered in general development plans; new environmental sacrifices must be avoided."[98] Above all, the leaders stress water conservation and reforestation. Zeng Peiyan, former SDPC chairman and head of the government office in charge of the campaign, suggested that Beijing would offer grain from "overflowing warehouses" to encourage peasants to abandon farming on marginal land and to plant trees. Already, local governments are drafting their plans to convert cropland to forest and grassland; Xinjiang, for example, plans to plant trees on 200,000 hectares of cropland, and Shaanxi plans to invest 20 billion yuan ($2.5 billion) to restore "the local ecological balance."[99]

Despite leaders' repeated assurances that environmental protection is a top priority in the campaign, many Chinese environmentalists have doubts. On March 7, 2000, *China Environment News* published several articles critical of the "Western development craze." One article pointed out that the west suffers from "widespread soil erosion, low agricultural productivity, water shortages and water quality problems," noting further that "[i]rrational development could cause significant ecological damage."[100] In an interview with reporters, Xiao Zhouji, professor of economics at Beijing University, called for caution when developing the west: "The development of the West must completely respect the protection of natural resources and the environment. The illegal logging and development of forestland is a constant occurrence. Soil erosion is serious, and desertification is intensifying. The plundering and exploitative nature of mining is a very deep lesson. Therefore as we develop the West, we must pay attention to protection of the environment, protecting the resources, maintaining the ecological balance, and promoting sustainable development."[101] A 2001 conference on the role of the nonstate sector in the development of China's west also was noteworthy for the skepticism voiced. One researcher from the Sichuan Academy of Social Sciences complained that Beijing had failed to consult scholars from the western region and described the campaign as "Western Exploitation, Eastern Development." She also feared that any real attempt at ecological reconstruction would harm those already struggling, noting that in Yunnan Province, the four million farmers who had restored their land to forest and grass had now "relapsed into poverty."[102]

Attempting to assuage the fears of environmentalists, the central government announced in 2003 that a substantial portion of its new investment in the west—almost $60 billion—would be devoted to environmental protection, including forest protection, reclaiming farmlands for forests, combating desertification, and closing pastures for the renewal of grasslands.[103]

Yet many in the environmental community remain skeptical. Liao Xiaoyi, president of the NGO Global Village of Beijing, for example, commented in response to the government's 2003 announcement,

> Investing in the environment is a good idea, but how much will be invested and what are the guarantees that this money won't be misspent or pilfered by corrupt local officials? Did the State Council consult non-government organizations like us for proper and constructive input? No. Just budgeting this amount of money isn't enough. The government must also come up with an auditing system to make sure this money will be spent properly.[104]

Not stated publicly but widely acknowledged privately is the concern among many environmentalists that SEPA was not included among the twenty-two–agency leading group charged with developing the Go West campaign. Without the direct input of SEPA at every stage of the planning process, many fear that the necessary comprehensive approach—integrating appropriate environmental, conservation, and efficiency technologies with development plans—will not occur.

The potential of the Go West campaign to exacerbate an already serious environmental problem is greatly magnified by the explicit linkage between the campaign and national security. In some ways reminiscent of the Qin dynasty's efforts to unify the country through major public works projects, China's leaders today view the Go West campaign as essential to their solidification of control over border regions such as Tibet and Xinjiang. These regions have the highest proportion of non-Han minorities in the country,[105] including twenty million Muslims as well as Tibetans and other ethnic groups. (Throughout China, the Han are the dominant ethnic group in the country equaling almost 92 percent of the population.[106] In Xinjiang and Tibet, however, despite a massive influx of Han Chinese since 1949, Uighurs and Tibetans outnumber Han in both regions.)

As one report noted, the campaign is the "fundamental guarantee for us to foster national unity, to maintain social stability, and to consolidate the borders."[107] General Chi Haotian, in a ten-day swing through the region in May 2001, discussed the vital role of the development drive in "consolidating national defense and realizing the country's long-term security and stability."[108] In meetings with party officials from Tibet, then Vice President Hu Jintao also stressed the importance of "stability" as a "premise" for economic development, noting,

> In the new situation, it is inevitably required by the stability of the overall situation of the whole country to maintain the stability of Tibet. Herein lies the fundamental interest of the people of all nationalities in Tibet and the guarantee for Tibet's development. We should unswervingly safeguard unification of the motherland and oppose separation with a clear-cut stand.[109]

Such obvious plans to "colonize" the region further have already become problematic for Beijing. A joint Chinese–World Bank plan to resettle 58,000 Chinese farmers on fertile lands historically inhabited by Tibetans drew international opposition from those who feared that the ethnic Chinese would swamp the culture and autonomy of the resident Tibetans.[110] After initially approving the resettlement loan, the World Bank later withdrew its support when an independent review committee criticized the plan for failing to protect ethnic minorities and subjecting them to a "climate of fear" by not guaranteeing confidentiality when consulting their views on the project.[111] Xinjiang is another area of concern for the leadership. It is the site of frequent bombings, protests, and even riots by separatists and independence advocates. Since the mid-1990s, the Chinese government has repeatedly arrested and executed those in Xinjiang suspected of fostering separatism either through armed or intellectual pursuits.[112]

The Go West campaign also raises concerns for the future of the environment in Tibet. According to at least one report, much of Tibet's environmental degradation appears to have come at the hands of Han settlers, including rampant deforestation—up to 40 percent of Tibet's old-growth tropical and subtropical mountain forests have already been clear-cut and shipped east. Rare plant and animal species are becoming extinct as Chinese officials exploit them for foreign markets, and the grasslands are suffering as Chinese demand for meat has led to overstocking of yak, goat, and sheep and the loss of Tibetan traditional herding and grazing practices.[113]

As multinationals examine their business prospects in China's west, they are cognizant of the political, social, and environmental landmines that await them. For some, the potential cost far outweighs the potential gain. Others, however, have attempted to work with their Chinese counterparts—in some cases dragging them along—to mitigate these costs by addressing them up front. Royal Dutch Shell took such a tack. At the outset of its discussions with PetroChina to help develop the West-East pipeline, Shell insisted on conducting its own environmental impact and social impact assessments.

Shell's environmental history in China is mixed. Over the past several years, the company has begun to make a name for itself by supporting a number of environmental NGOs in China, including Friends of Nature, Global Village of Beijing, and Friends of Green in Tianjin. At the same time, it has locked horns with environmental activists in Hong Kong over a massive joint venture petrochemical plant in Huizhou near Daya Bay in southern China. Of greatest concern was the potential for significant water pollution and harm to marine life, although Chinese officials involved in the project insist that only the most advanced environmental protection technology would be utilized at extra cost to the project. Resettlement of the thousands of villagers required to make room for the plant also has been complicated both by corruption in the delivery of compensation and by migrant workers who want the same benefits as the permanent residents.[114]

Nonetheless, Shell officials took great pains to ensure that they avoid such problems with the pipeline project. This despite significant risk that the pipeline would prove to be an economic sinkhole for the company; there was not yet an infrastructure, distribution system, or market for natural gas for much of the pipeline's designated destinations.

Although Shell's Chinese partner, PetroChina, had ostensibly undertaken an environmental impact assessment before initiating work on the pipeline, Shell officials believed that more needed to be done. As the lead oil company in the foreign consortium that would have owned 45 percent of the project, Shell was aggressive in its efforts to ensure that environmental and social considerations were factored into the business plan, as well as into the actual laying of the pipeline.

On the environmental front, Shell hired the well-respected environmental consulting firm Environmental Resources Management (ERM) to manage the impact assessment. In addition, ERM placed twelve to fifteen teams of people along the pipeline route to ensure that environmental challenges were addressed before it was too late. In one instance, for example, PetroChina, which began laying pipeline even before the investment deal was realized, stopped construction when an ERM team pointed out that it would necessitate knocking down part of the Great Wall. Other sites of cultural, historic, and environmental value were also preserved, such as endangered monkey habitats in central China and a swath of land traversed by rare camels.

In addition, Shell officials were concerned about the social impact of the pipeline. Although only two hundred or so families would need to be relocated, Shell officials hired the UNDP to undertake an exhaustive socioeconomic baseline survey, interviewing communities along the pipeline route to determine their welfare interests and needs. This went beyond the demands of the Chinese government and is modeled on World Bank practices. While distribution of revenue would have been determined by the central government, Shell also insisted that once the pipeline was operational, part of the revenues would be devoted to community development needs such as poverty alleviation and education. This would have been especially important in Xinjiang, where a perception that Beijing and the international community would have benefited from Xinjiang's resources could have presented political problems and added fuel to ethnic protests and potentially violence among the Uighur population.

Despite Shell's efforts, the Free Tibet campaign continued to challenge and criticize the development of the pipeline and Shell's participation in the project. With its emphasis on rights for minorities, the campaign was not likely to sanction any Beijing-based initiative with a multinational that exploited the resources of minority areas. Still, Shell, along with its international partners, Exxon Mobil and Gasprom, moved forward with the project, until negotiations collapsed in mid-2004.

As the Chinese leadership presses forward with the Go West campaign, therefore, much may depend on the role of international actors to ensure that the government fulfills its pledge to develop the region with sound conservation policies. At stake are the credibility of China's commitment to environmental protection and the long-term future of China's western regions. Certainly the government confronts skepticism both at home and abroad on this front. In the meantime, Chinese scholars and environmental officials, as well as international nongovernmental organizations, are certain to watch carefully the involvement of international governmental organizations, multinationals, and foreign governments.

A Transforming China

In the three decades since the 1972 UN Conference on the Human Environment, the world has witnessed a sea change in Chinese environmental protection attitudes and practices. The country has been transformed from one with no environmental protection apparatus, no environmental legal system, and only the smallest environmentally-educated elite, to one in which numerous bureaucracies are engaged in protecting the environment, the legal infrastructure embraces virtually every aspect of the environment, and there is a vast, ongoing environmental education effort throughout the society.

The international community has played an essential role in this transformation. China's participation in international environmental regimes has contributed to the development of a domestic environmental community, environmental laws, and new bureaucratic arrangements to manage environmental protection. These new actors and arrangements, in turn, have provided new opportunities for international environmental cooperation. Over the years, this synergy has produced an enormous range of cooperative ventures involving every sector of Chinese society and level of Chinese government, as well as the full range of international actors.

Moreover, the modernization of China's economy now invites an entirely new level of cooperation, involving innovative policy approaches, advanced technologies, and greater technical expertise. And the country's increasing involvement in international trade regimes holds promise for reinforcing the integration of environmental objectives in China's future economic development.

But the need for a strong enforcement capacity as well as an incentive system for environmental protection remains a significant obstacle to the implementation of many new market-based environmental policies and advanced technologies. While some cooperative ventures have succeeded despite these impediments, many have failed to achieve their full potential. Moving international environmental cooperation forward will depend heavily on the ability of the Chinese leadership to make progress in these two areas of enforcement and incentives. In addition, the continued penchant of China's leaders for large-scale, environmentally degrading development campaigns not only draws into question their commitment to environmental protection but also opens the door to substantial international, and increasingly domestic, criticism of both China and the country's international partners.

Future cooperation on environmental protection between China and the international community holds as many challenges as opportunities. The interest of the international community and proactive actors within China in advancing such collaboration and significantly enhancing China's environmental protection capacity is clear. What remains is for China's leaders to invest the necessary political and financial capital to develop an effective incentive system in which environmental best practices will thrive.

China is not alone in the particular blend of policy measures it has adopted to respond to its environmental challenges. The former states of Eastern Europe, republics of the Soviet Union, and several Asian countries shared many of the economic, political, and institutional challenges to effective environmental protection that China currently faces. Moreover, the boldest policy choices of these countries to protect the environment—to open the door to NGOs and greater public participation, as well as deeper integration into the international community—are the same ones China is now pursuing. In so doing, these countries became vulnerable to far greater challenges to their systems of governance and seismic shifts in the nature of their polities. In chapter 7, we'll consider the implications for China of these lessons from other countries.

CHAPTER SEVEN

LESSONS FROM ABROAD

M any countries face challenges similar to those of China in balancing environmental and economic demands while maintaining social stability. The scale of China's environmental degradation and pollution, however, dwarfs that of most countries. China's economic and political transition further complicates efforts to compare China with other countries and draw useful lessons from their experiences. Nonetheless, there are broad similarities between China's current set of challenges and that confronted by other Asian countries, as well as several countries of the former Eastern Europe and republics of the former Soviet Union. Taken together, these countries reflect the range of China's environmental problems, as well as the political and economic dynamics of China's transition from a totalitarian to an authoritarian regime and a command to a market economy. The recent history and contemporary situation of these countries thus help provide a forecast of potential changes and trends in China's environmental protection efforts.

More than a decade ago, the Eastern Bloc countries and republics of the former Soviet Union confronted the challenge of rapid industrialization and massive environmental degradation within the framework of socialism, much as China has for the past several decades. Their economies were dominated by highly polluting state-owned enterprises (SOEs); they treated natural resources as free goods; and they paid little attention to energy consumption or pollution control.[1] The leaders pursued large-scale development projects, including dams, river diversions, and agricultural experiments, with devastating environmental consequences. Environmental protection was of little real concern.

Yet in critical ways, China today presents a very different picture from Eastern Europe and the former Soviet Union in the late 1980s and early 1990s. China has moved much further along the path to a market economy and integration into the global economy. The Chinese leadership continues to shed parts of its communist system, privatizing the economy, breaking free of SOEs, developing land tenure laws, pushing toward a free labor and genuine housing market, and in the process, permitting large-scale economic inequalities to emerge. While continuing to cloak its behavior in claims of social justice, the Chinese Communist Party (CCP) has become largely devoid of ideological content, serving primarily as a patronage machine committed to rapid economic development.

In this way, China has begun to resemble more closely the countries of Central and Eastern Europe today, as well as its neighbors to the south and east—the economically free-wheeling but politically constrained authoritarian states of Asia. At various points in their recent history, Korea, Thailand, Malaysia, Indonesia, and the Philippines have shared with China several of the daunting challenges inherent in balancing environmental protection with rapid economic development. In addition, these authoritarian states lacked, and in some cases still lack, the transparency, accountability, and political institutions necessary for effective environmental protection.

In confronting their environmental challenges, these other countries' initial decisions favored redressing environmental degradation through a combination of political and institutional reform rather than by increasing investment in environmental protection or restructuring the economy to encourage economic actors to integrate environmental protection into their development plans. Thus, they began by developing an environmental bureaucracy and legal system. Like China, they later opened the door to greater public participation in environmental protection as a means of both addressing the populace's growing environmental concerns and releasing the broader pent-up pressures for political change. The international community also played an important role in shaping the environmental situation in many of these countries. In the end, this approach affected not only the evolution of the countries' environmental situations but also their broader political economies.

Eastern Europe and the Soviet Union

Rapid industrialization after the Second World War brought a host of environmental problems to Eastern Europe and the former Soviet Union, many of which resemble the challenges facing contemporary China. Into the 1980s, much of the region relied exclusively on coal for its energy; in the German Democratic Republic (GDR) and Czechoslovakia, for example, more than 80 percent of the countries' energy needs were met by coal. As a result, Czechoslovakia produced more sulfur dioxide than any other country in central Europe,[2] with harmful consequences for agricultural productivity, plant life, and human health. Bohemia registered shocking levels of respiratory disease, digestive ailments, heart problems, and birth defects from its coal-based air pollution.[3] East German scientists also acknowledged that the incidence of heart and respiratory diseases as well as cancer were 10 percent to 15 percent higher in highly polluted regions of the country. In addition, mining and coal use contributed to the loss of agricultural land, falling groundwater levels, and increasing rates of water pollution from sulfates, phenols, sulfuric acid, and particulate manner.[4] Importing clean coal technology, such as desulfurization equipment from the West, was viewed as too expensive.[5]

Poland and Hungary also posted frightening statistics with regard to water quality. By the late 1980s, Polish officials believed that only 1 percent of their surface water was safe to drink, and as much as 49 percent was unfit even for industrial use. Moreover, almost half of all major Polish towns had no sewage treatment plants and dumped their wastewater untreated into rivers and lakes.[6] In Hungary, two-thirds of the sewage water remained biologically unpurified.[7]

The negative impact of development on the environment was compounded by the nature of the political system. Environmental laws were not enforced, managers were not rewarded for environmentally sound production but solely the quantity of goods produced, access to information was limited, and the right to organize politically was restricted.[8] In Poland, for example, in 1983, the Council of Ministers cited twenty-seven environmentally endangered areas and put into place regulations to prevent the creation of new industries harmful to the environment.[9] Yet there was virtually no enforcement of these regulations. This system of political constraints and perverse economic incentives produced shocking levels of pollution and related social costs.

First Steps: Enhancing Government Awareness

As in China, the governments of Eastern Europe began to acknowledge the challenge of environmental protection during the 1970s and 1980s, particularly in the aftermath of the 1972 United Nations Conference on the Human Environment (UNCHE). During this time, many of these countries established environmental protection agencies,[10] and officials began to acknowledge publicly the interdependence between economic growth and environmental protection. In East Germany, the mantra became "an efficient economy assures adequate investment funds for environmental protection."[11] Numerous environmental laws were enacted.[12] By the late 1980s, for example, Poland had passed thirty environmental laws and resolutions; Czechoslovakia had adopted fifty-three laws and regulations; and Yugoslavia had issued more than three hundred. East Germany, Romania, and Hungary all passed comprehensive state environmental protection laws during the 1970s, and Bulgaria enacted a set of revised environmental management guidelines.[13]

Yet the impact of these efforts was limited. As two analysts have commented, "It was one of the greater paradoxes of the communist period that governments were prolix in legislation aimed at limiting environmental damage, yet this legislation did very little to halt the degradation."[14] Corruption impeded implementation: "Past experience suggests that money voted for environmental protection is often diverted to other purposes by local people's committees and that exemptions to the water pollution laws are often granted on a wholesale basis."[15] The primacy of economic development was also a constraining factor: "Managers were rewarded for fulfilling their plans and severely penalised for not doing so. . . . The result was considerable waste, notably of energy resources, and with it pollution on a massive scale."[16] As in China, too, outdated factory equipment, low fines for polluters, and environmental inspectors who were overruled on economic grounds all contributed to poor implementation.[17]

At the same time, the governments also engaged in efforts—which presaged those of China during the mid-1990s—to raise the profile of environmental protection in their countries. Many countries doubled their investment in environmental protection in their 1986–1990 plans, and environmental impact assessments were enacted. There was recognition in Poland, for example, that environmental degradation was a barrier to further socioeconomic development, and environmental protection was rated as highly as housing, food, and energy in public opinion polls. In Yugoslavia, banks were required to account for environmental demands when granting loans for new projects.[18]

Finally, the governments sponsored official environmental protection organizations devoted to scientific assessment of environmental degradation and public campaigns to clean up the environment. Most often, they were also focused on issues of conservation. For example, in Czechoslovakia during the 1980s, the Slovak Association of Guardians of Nature and the Homeland used youth volunteers to clear up polluted regions and to plant trees.[19] Together with its Czech counterpart, it boasted almost 30,000 individual members in the early 1980s. In Bulgaria, in the early 1980s, the journalists' union and the National Committee for the Protection of Nature jointly agreed to cooperate on environmental protection measures.[20] In the former GDR, by 1986, nearly 60,000 people had joined the officially sponsored Society for Nature and the Environment.[21]

These state-sponsored groups did not, however, develop a broader or more systemic critique of environmental protection. Constraints included poor access to reliable data, severe limits on the activities they could undertake, and limited interactions with foreigners.[22]

East European scientists also began to focus on environmental issues. In Hungary, scientists were among the leaders to initiate environmental protection. They focused their attention on the environment at the 1971 Annual Assembly of the Hungarian Academy of Sciences, and in 1972 held a special conference on the topic.[23] In the mid-1980s, the Polish Academy of Sciences recommended that state investment in environmental protection for the 1986–1990 five-year plan ought to be increased by 6 percent.[24] These scientific and environmental experts were the "most visible and authoritative spokespersons for the environment," forming a community that crossed professional boundaries and included legal experts, scientists, and social scientists who knew each other from their universities or by virtue of their involvement in one of the nascent environmental organizations.[25]

The Next Stage: Environmental Activism

Yet these state-sponsored efforts failed to address the underlying problems—both environmental and political—that plagued the region. What emerged to fill the vacuum were environmental clubs with broad societal representation that responded aggressively to the overwhelming pollution problems.

Non–state-organized environmental activism in Eastern Europe dates as far back as the establishment of the Polish Ecology Club in September 1980.[26] The development of the club was spurred by concern throughout the country about threats to health within heavily polluted areas such as Silesia, and there was broad public support for reviving the long-held Polish tradition of maintaining national parks and nature reserves.[27] Polish scientists, cultural elites, trade union members, and environmentalists banded together to call on the local Kraków government either to close or to modernize several of the major factories in the city's vicinity. With the support of the Kraków mayor, they achieved some success, and by July 1981, the Polish Ecology Club had fourteen branches and twenty thousand members.[28]

In Hungary, as early as the early 1970s, citizens' groups petitioned Budapest concerning air pollution and unbearable noise. In one case, in 1977, citizen pressure resulted in the closure of a polluting factory after children became ill from lead poisoning.[29]

In the GDR, the Protestant Church led the fight for environmental protection. In the 1970s, the church promoted environmental discussion groups and activism among parishioners. According to one estimate, at least several hundred people were involved in such unofficial groups, which held "environmental worship services" in heavily polluted areas.[30] While much of their work was devoted to activities such as tree planting or riding bicycles, they also protested a chemical works plant and sponsored discussions on nuclear power.[31] The Lutheran Church even published a clandestine environmental journal, *Umweltblatter* (Environmental Pages).[32]

The Catalyst: Chernobyl and Its Aftermath

In April 1986, the nuclear disaster at the Chernobyl Nuclear Power Station in the republic of Ukraine provoked a radical change in Soviet politics, with effects that rippled not only through the republics of the Soviet Union but also through the satellite nations of Eastern Europe. The now infamous explosions at Chernobyl on April 26 emitted more radioactive material into the atmosphere than the atomic bombs that leveled Hiroshima and Nagasaki.[33] The radioactive gas and particles dispersed as far as Greece, Yugoslavia, Poland, Sweden, and Germany. The Soviet leadership, however, did not report the accident to the public until two full days after the accident, and even then attempted to minimize the disaster by claiming that "the radiation situation in the power station and the surrounding area is stabilized."[34]

The domestic and international outrage at the way the disaster was managed proved a spur to openness in the Soviet Union and emboldened the populace to challenge the government's policies on other environmental issues. Throughout the Soviet Union, intellectuals, students, and scientists led the drive to establish ecology clubs that pressed officials to address local pollution problems. In the face of massive protests in 1986 against the Siberian River diversion project, which was to provide water to the parched cotton and rice fields of Central Asia, for example, the government canceled the project, which had been decades in the making.[35]

By the late 1980s, environmentalists in the Soviet Union were having a noticeable impact on the path of economic development. Almost two-thirds of the thirty-six projects planned by the Ministry of the Medical and Microbiological Industry—reputed to be a serious polluter—had to be canceled when local groups refused to grant the necessary land.[36] In 1989, more than one thousand factories were shut down or had their operations scaled back as a result of environmental violations.[37] Thus, as Murray Feshbach and Alfred Friendly conclude, the "passage from pollution to politics was an easy one."[38]

Simultaneously, pressures were building in Eastern Europe for both enhanced environmental protection efforts and greater public input into governmental policy making. In Poland, many environmental groups attributed both Chernobyl and problems in the Polish environment to the lack of a public voice in the decision-making process.[39] The media played a crucial role in opening up the political space for public debate and discussion. As East European analyst Christine Zvosec argued at the time, "This is due not to any liberalization in reporting policy, but to the inability of ignoring pollution problems."[40] Environmental problems provoked media criticism of unusual frankness. In Poland and Czechoslovakia, the media referred to the respective environmental situations in each country as an ecological catastrophe or disaster.[41] In Yugoslavia, for example, the press extensively publicized the efforts of environmental experts to prevent the construction of the Tara River Dam, and a Belgrade radio program not only provided information on pollution but also exposed environmental polluters.[42]

The Third Stage: Beyond the Environment

For the governments of the Soviet Union and Eastern Europe, environmental protest appeared, at least initially, as a "relatively safe outlet for expressions of more general discontent."[43] As one Russian environmentalist stated, "Green activities were tolerated because officials did not at first identify them as threats to Communist rule."[44] From the perspective of government officials, properly managed, popular participation in environmental protection could be constructive:

> Criticism should not be superficial or one-sided, but rather sympathetic to the dilemmas facing decision makers. Environmentalists should be willing to seek realistic solutions and to propose alternatives. Apart from criticizing existing government policy and warning about potential hazards, social organizations can also play an important role in increasing the public's awareness of environmental questions. Active cooperation between governmental and nongovernmental organizations is a precondition for effective environmental protection.[45]

But the ecology groups quickly disabused these officials of their belief in the relatively nonthreatening nature of their activities. Activists who began by fighting pollution became leaders in the broader battle for "cultural, economic and political independence."[46] In Latvia, for example, the first "environmental" protests were in fact cultural in nature: In 1984, the activists were devoted to restoring churches and monuments.[47] The Latvian Environmental Protection Club (*Vides Aizsardzibas Klubs;* VAK) organized activities devoted to reinstating the Latvian language, commemorating the Latvian flag, and honoring the Latvians deported by Stalin in 1940.[48] As one Bulgarian activist noted, "Ecological activity was the only permitted form of action. If you acted around human rights or religious freedom you just went to prison. Environmentalism was the only way to express civil disobedience without being arrested."[49] Consequently, VAK served as a point of coalescence for activists of all stripes. In an interview with Jeffrey Glueck, Valdis Abols, a VAK vice president, stated, "Many, maybe even the majority of people who joined us in 1987, were very far from green thinking. They just saw that this was a political force. They could use this for their political ambitions."[50] Similarly, a Slovak activist stated that environmental protection organizations were "the only game in town." This meant that activists interested in other issues, for example assisting the rights of the disabled, joined environmental protection organizations because they were the only nongovernmental organizations (NGOs) legally sanctioned.[51]

Other social organizations incorporated environmental issues into their platforms. In Czechoslovakia, for example, in 1983, Charter 77, an association of human rights activists founded in 1977, published information from a report by the Czechoslovak Academy of Science that had been denied publication by the Czech authorities.[52] Charter 77 often discussed environmental issues in its publications, and in 1985, two environmentalists were sentenced to jail for drawing attention to the deterioration of the forests in Bohemia and publicly criticizing the Czech government's environmental protection efforts.[53] This did not deter the organization; in April 1987, Charter 77 published a full-length treatise on the environment that called for restructuring the economy and elevating the environment to a first priority issue for the government.[54]

In Poland, the labor organization Solidarity also made the environment part of its political platform. In May 1981, for example, Solidarity's Interfactory Workers Committee advanced a resolution calling for work to stop at an extraction area until environmental damage could be addressed.[55] By 1988, there was a Green party in Poland.

The linkage between the environment and broader political issues such as nationalism and anticommunism, was often made explicit through opposition to large projects such as nuclear power stations, dams, and river diversions.[56] Severe problems of transboundary pollution and other environmentally related disputes also fostered environmentalism in Eastern Europe. For example, in 1987, Polish students demonstrated against the accidental discharge of heating oil into a tributary of the Oder River by a Czech factory. When the

Czech government did not report the incident for several days, the Polish government complained about the lack of information and demanded compensation.[57] Similar conflicts arose between Austria and Hungary. As Miklós Persányi explains,

> A very heated nationwide debate developed over the importation of waste from Austria. An enterprise supported by the local council in Mosonmagyaróvár in western Hungary contracted to dump garbage sent from Graz. The Austrian partner paid with hard currency and promised to provide high technology waste management equipment. Chemists from a local environmental group analyzed the waste and determined that it contained heavy metals and that its pollutants were reaching groundwater. The group informed the public and the authorities, leading the National Authority for Environmental Protection and Nature Conservation to stop the importation. Subsequently, a regulation prohibiting the importation of hazardous wastes was adopted.[58]

Two of the most compelling cases were the conflict over the treaty between Czechoslovakia and Hungary to build the Gabcikovo-Nagymaros Dam on the Danube with construction by Austrian firms and the transboundary air pollution dispute between Romania and Bulgaria.

In the former case, the debate began when leading Hungarian environmentalist Janos Vargha argued in a 1981 article that the dam would "change the hydraulic, physical, chemical, and biological conditions of a nearly 200 kilometer section of the river itself and also that of the surrounding ground water. These changes would be harmful to the drinking water supply, the quality of the river and ground water, agriculture, forests, and fish, as well as the picturesque landscape."[59] By 1984, a grassroots environmental organization, the Danube Committee, had been established to oppose the dam. This organization later grew to encompass other environmental concerns as well as anticommunism, becoming known as the Danube Circle, which was registered as an official organization in 1988 and soon began to work with groups in Austria and other international NGOs. At the same time, Hungarian environmentalists began to protest Hungary's agreement to dispose of toxic waste from Austria.[60] In 1988, forty thousand people demonstrated in Budapest against the construction of the dam on the Danube.[61] One year later, the government of Hungary decided to halt construction of the dam, while the Czechs continued its development. However, both domestic NGOs and international NGOs became active in pressing the government to cease construction. The International Coalition Against Large Dams and the International Rivers Network (both of which have agitated against China's Three Gorges Dam) also pressed international investors and the government to stop the development of the dam.[62]

Even in Bulgaria, one of the most repressive of Eastern European societies, the issue of transboundary pollution awakened an environmental spirit within the populace. In 1987, the Bulgarian town of Ruse recorded dangerously high levels of chlorine pollution, sparking a series of demonstrations that fall. The Bulgarian government was forced to raise the issue with Romanian officials.[63] This became the first real challenge to the Communist Party. The Committee for the Defence of Ruse, which was established to address the issue of the chlorine pollution, later gave rise to the broader Ecoglasnost, which then became a core component of the Union of Democratic Forces.[64] Although the official bilateral negotiations successfully resulted in Romania shutting down more than 80 percent of the factories, environmental activism was unable to emerge under President Nicolae Ceausescu, despite terrible circumstances.[65]

Environmental issues also served as a catalyst for political change in several of the Soviet republics. As D. J. Peterson points out, "To many citizens, the destruction of nature in their homelands epitomized everything that was wrong with Soviet development, the Soviet economy, and the Soviet state itself, and these great injustices against nature were obvious and easy focuses for action. Nature became a medium for social change."[66] In 1989, the Lithuanian greens argued that "besides a lack of pollution control equipment, the republic suffers 'a lack of control over its production and resources' . . . [and] accuse[d] Soviet occupiers of turning the republic into 'a colonial industrial dump site producing goods and services far beyond the needs of its own inhabitants' "[67] In Georgia, in 1988, popular protest against the Transcaucasus Main Railway laid the foundation for the creation of the Ecology Association under the auspices of the All-Georgia Rustaveli Society. This became the forerunner of the movement that asserted Georgian independence.[68]

Finally, "there were those for whom environmental groups were merely a vehicle for pursuing the main goal of overthrow of the Communist regime. . . . This was to be especially so in countries where environmental interests were not deeply rooted in the value system of the people, such as Romania."[69] In such cases, these "environmentalists" might do little to enhance environmental protection, electing instead to focus their energies on more politically salient issues.

Romania, for example, had one of the most polluted cities in Eastern Europe, Copsa Mica, where the people were routinely covered in black carbon powder emitted from a local tire factory. Within the town, more than half the almost three thousand people tested showed symptoms of lead poisoning. During 1983–1993, about two thousand people were hospitalized for severe lung and stomach pains or anemia due to the lead poisoning. Ninety-six percent of the children between ages two and fourteen had chronic bronchitis and respiratory problems.[70] Yet, the local people were so dependent on the factory for their livelihood, they perceived the pollution as "part of reality"[71] and were reluctant to agitate. Moreover, despite promises from the United Nations that it would provide financial support for improving the situation, Romanian environmental NGOs, such as the Ecological Movement of Romania, made no push for change. They were more concerned with blocking Ceausescu in his efforts to turn the Danube delta into agricultural land plots because this issue was directly linked to the dictator's legacy.[72]

Postscript: A Decade Down the Road

The political and economic transition of Eastern Europe and the former Soviet Union has produced some environmental success stories. In some of the wealthier countries, such as the Czech Republic, Poland, and Hungary for example, a combination of economic reform, public pressure, and environmental activism has produced significant declines in air, water, and soil pollution.[73] These countries have greater resources to invest in cleaner, more efficient technologies, and have moved away from the pollution-intensive industrial practices of their communist past toward more efficient production methods.[74]

Significantly, the desire of many of these countries to join the European Union (EU), which is imposing stringent environmental standards for prospective members, is forcing the governments to ratchet up significantly their environmental protection efforts. Poland, for example, has achieved some notable progress in its environmental record driven in part by EU mandates, taking tough steps to ban the import of waste from other countries, to increase the fees for environmental pollution, and to experiment with market-oriented measures such as tradable permits. In Hungary, some believe that the government's desire to accede to the EU, rather than the actions of Hungary's flourishing environmental movement, is the key to understanding the country's efforts to improve its environmental record.[75] Poland, Hungary, and the Czech Republic were all invited to join the EU on May 1, 2004.

Yet in other countries in the region, the promise of a greener future heralded by the rise of environmental activists and NGOs and the downfall of communism has yet to produce significant improvement in the environment. There are several reasons for this. Primary among them is the continued economic straits in which many countries find themselves. The case of Bulgaria, for example, suggests that the government's perceived ability to invest in environmental protection has varied according to its overall economic situation; thus during the mid-1990s, as the economy continued to deteriorate, Bulgaria's investment in environmental protection dropped from 1.3 percent of gross domestic product (GDP) in 1993 to 0.9 percent in 1995; Slovenia similarly witnessed an overall economic decline and corresponding decrease in investment in environmental protection.[76]

Even as the countries of east and central Europe continue to negotiate the legacy of their communist years such as aging nuclear facilities and highly polluting heavy industrial factories, rapid economic growth has provided an unsettling array of new environmental problems similar to those of China: growing air pollution from the increasing number of cars on the roads; increasing contamination of the water supply from new pollutants, such as detergents; and substantial increase in problems of waste management from plastic packaging.[77] In many of these countries, environmental NGOs have had difficulty transitioning from broader social activism to the nuts and bolts of shaping environmental policy to resolve such new challenges.

Political considerations, too, often constrain the opportunities for these NGOs. Senior Russian officials, for example, have attempted to prevent environmentalists from reporting on ecological damage by Russia's military by framing the issue of nuclear waste as a security issue and accusing the activists of treason. Russian president Vladimir Putin, with support from other Russian officials, also successfully blocked a national referendum that garnered more than two million signatures opposing the import of nuclear waste, a program that would earn Russia more than $20 billion over ten years.[78] Indeed, the Russian government's evident reluctance to respect the role of popular involvement in environmental affairs has led some environmental activists to link environmental protection with democracy once again: "[The referendum] is not just a fight against nuclear waste import, but a fight for establishing democracy and strong civil society in Russia."[79]

With few exceptions, too, the environmental protection agencies throughout the central and east European states remain weak, particularly in the face of continued demands to raise the population's standard of living. In the case of Russia, President Putin dissolved Russia's State Committee on the Environment in 2000, further hampering the work of Russia's environmentalists, particularly since environmental oversight was then passed to the Natural Resources Ministry, the primary agency in charge of mining, oil exploration, and timber extraction.

Finally, many environmental NGOs remain elite-oriented organizations with stronger ties to international funding institutions than to their domestic constituents. Grassroots organizations, in contrast, may be more effective in reaching their constituents but often lack the funds or organizational skills to sustain them over the long term.

Asia Pacific: From Economic Miracle to Environmental Catastrophe

At first glance, the Asia Pacific countries appear to have little in common with those of Eastern Europe and the former Soviet Union. From the 1980s until the onset of the Asian financial crisis in the late 1990s, the Asia Pacific region enjoyed unparalleled economic growth. Economies grew at unheard of rates, producing unprecedented levels of affluence throughout the region.

Yet this rapid development also produced an environmental disaster that was largely ignored in the planning calculus of the region's leaders. Like the countries of Eastern Europe and the former Soviet republics, the Asia Pacific region as a whole suffered from disregard for environmental protection, weak environmental protection institutions, and little opportunity for public participation. Moreover, like China, the Asia Pacific region also suffers the environmental burdens of growing automobile use, urbanization, and migration, among other concerns.

Rapid economic growth in the Asia Pacific has exerted a number of negative environmental impacts. Water pollution is the chief problem. Factories dump their untreated domestic waste directly into streams, rivers, and coastal waters. Throughout Southeast Asia, denuded uplands have contributed to devastating cycles of flooding and drought. Agricultural runoff, untreated domestic sewage, and industrial waste also degrade ground, surface, and coastal waters. By the early 1990s, World Bank estimates indicated that in Indonesia, the Philippines, and Thailand, most pollutants were growing three to five times faster than the underlying economies. In metro Manila and several other cities in the Philippines, less than 20 percent of households can access direct water supplies, and only 14 percent of the population benefits from sewage treatment.[80] In fact, sewage treatment is nonexistent in most cities in the region, and according to a mid-1990s World Bank report, groundwater was polluted by "cesspools, septic tanks, leaking sewers and landfill sites."[81] Moreover, in Thailand, Vietnam, and Indonesia, so much water was being "sucked up from wells" that aquifers were being polluted by seawater.[82]

Poor water quality and water scarcity have threatened both the health of the region's populace and its continued economic development. In the Java Sea, increasing levels of waste, including toxic waste from industrial and processing activities, are causing declining fish stocks and increased morbidity among the human population from the spread of infectious diseases. In 1994, in Vietnam and Hong Kong, the inadequacy of water treatment facilities was blamed for outbreaks of cholera.

Even in such wealthy nations as Korea and Taiwan, industrialization threatens water quality. In Taiwan, high levels of cadmium and copper have contaminated several major waterways used for drinking, aquaculture, and agriculture.[83] In Korea, untreated industrial waste has polluted several primary sources of drinking water.[84] While Japan has successfully reduced levels of mercury and arsenic in its coastal waters, the development of coastal resorts and golf courses has devastated fish stocks. Overall, degraded water resources hamper continued economic growth throughout the region through increased morbidity and mortality rates, increased costs to industry, constrained supplies to agriculture, and decreased value of freshwater and coastal fisheries.

However, as in China, the most visible sign of environmental pollution in Asia may be the poor air quality. The pollution generated by automobiles, coal and oil use, and inefficiencies in the supply and distribution of energy have led to significant health and economic problems for the region. Air pollution has been growing two to three times faster than the economy, and the death toll from air pollution in Asia is alarmingly high. In Bangkok, for example, cars and motorcycles emit smoke and dust that contain lead and carbon, and doctors have found lead in the umbilical cords of newborn babies.[85] Indeed the air quality is so poor that in the mid-1990s, a group of environmental experts from the University of Hawaii who visited Bangkok to measure air pollution refused to return to the city. Air pollution is estimated to be responsible for one to two thousand deaths annually in both Bangkok and Jakarta.[86] In Manila, too, urban air pollution exceeds the limits set by the World Health Organization by 500 percent.[87] Japan, with its high levels of automobile use, has been unable to meet its own ambient air quality standards for nitrogen oxide in Tokyo and Osaka.

These problems are compounded by increasing population, migration, and urbanization in many of these countries. By 2015, it is estimated that almost two billion people, or more than half of the entire population of Asia, will live in cities.[88] The infrastructure necessary to meet basic environmental, population, energy, and sanitation needs such as sewage systems, housing, plumbing, power, and transportation is lagging far behind this growth.

The rapid and devastating manner in which the Asian Pacific countries have degraded their forest resources has also contributed to cycles of drought and flooding, soil erosion, and climate change. Many countries have derived a significant portion of their wealth from logging. The region boasts the highest average annual deforestation rate in the world (1.2 percent). Domestic needs for agricultural land and firewood for fuel are partly responsible for this deforestation. But as in China, the substantial income derived from wood-related products is the driving force behind much of the region's loss of forest cover. The process of economic development has also contributed to the loss of valuable arable and forested land. In Thailand, for example, Thai developers have turned fertile land over to developers of golf courses, resorts, and industry. Forested land has decreased from 53 percent of the total land area in 1961 to a mere 28 percent in 1988.[89] In the Philippines, forested land similarly declined by almost 50 percent during 1969 to 1993, transforming the Philippines from a major world exporter of wood in the 1980s to an importer by the 1990s.[90]

In addition to its large wood exporters, the Asia Pacific region is home to some of the most significant consumers of timber and wood products. While Japan maintains a 70 percent forest coverage and strictly regulates domestic timber-cutting, it is the largest net importer of hardwood in the world. In addition, Korea and Taiwan are also significant importers; since 1971, South Korea's imports have more than quadrupled, while Taiwan now imports over seven times its 1971 levels.

The Politics of Resource Exploitation

Despite growing economic affluence and the simultaneous decline in environmental health, political leaders in the Asia Pacific have been slow to rise to the challenge of environmental protection. Chief among the impediments to more effective environmental protection is the close financial, friendship, and often familial connection among elites responsible for environmental protection and those in the business communities. The importance of these ties is enhanced by weak legal infrastructures, especially in the economically less advanced Asian states.

In Malaysia, for example, leaders traditionally have held the right to allocate logging concessions to their political supporters. Despite measures to improve the situation, international NGOs, such as Friends of the Earth and Greenpeace, have stated that corruption is undermining these initiatives, noting that the political structure is closely intertwined with vested interests.[91] In Indonesia in 2003, experts blamed illegal logging for floods that caused the death of more than 100 people. The minister of environment stated at the time, "It is difficult to combat illegal logging because we must face financial backers and their shameless protectors both from the Indonesian armed forces and police and other government agencies."[92] This same pattern emerged in the Philippines, where loggers often bribed forest inspectors in order to log the timber illegally; in the Philippines, the illegal timber trade is valued at four times that of the legal trade.[93]

Weak environmental agencies also are a central problem for implementing successful environmental policies in virtually every Asia Pacific country. In part, this is because states typically established their environmental protection agencies later in the overall development of their bureaucracies or as spin-offs from other ministries whose primary commitment was not to environmental protection. Malaysia still does not have an independent environment agency; rather it is lumped together with scientific issues in a Science, Technology, and Environment Ministry. Typically, these environmental agencies are plagued with low levels of funding, poorly trained staff, inadequate equipment and technology, and mandates that overlap or conflict with those of other, more powerful agencies. In Thailand, for example, there are "about twenty government agencies with responsibilities related to the problems of water pollution, . . . many [of these] have made no commitment to enforce environmental regulations."[94]

Moreover, while environmental protection agencies may have regulatory power, most have no mandate to influence sectorwide development paths that are responsible for environmental degradation, such as energy or transportation. In Japan, the Environment Agency is generally considered among the most ineffective of all the government organizations. In 1993, for instance, when that agency attempted to advance a new Environmental Basic Law, the law was substantially diminished in scope by the resistance of various business and government officials, who were concerned that the law would elevate the status of the Environment Agency and hinder Japanese economic development.[95] In Korea, too, the Ministry of Environment did not assume control over monitoring tap water quality from the Ministry of Health and Social Affairs and the Construction Ministry until 1994, after the country experienced a high level of contamination at several key sources of drinking water. However, as the ministry had only two or three employees who possessed the qualifications to be classified as professional technicians, it turned over much of the management responsibility to the private sector.[96]

Even when national elites have supported stronger environmental protection efforts, regional authorities have often undermined them. In Malaysia, for example, regional authorities who received revenues from timber royalties have hampered the efforts of the national government to curtail illegal logging. Moreover, each state has autonomy over its forests and is responsible for issuing its own forestry ordinances, for zoning forest areas, and for instituting the mechanisms to enforce the laws. Yet in one heavily logged area—Sarawak—during the mid-1990s, there was only one forestry department official.

The reluctance of Asia Pacific leaders to take strong action by increasing investment in environmental protection, raising fees for polluters, or strengthening enforcement has left gaping holes in environmental protection efforts throughout the region. What has emerged to fill this gap in each country—with or without the active encouragement of the government—has been a hybrid of environmental protection efforts led by NGOs, the media, individual localities, and business leaders. This amalgam of approaches parallels that of China.

In a few cases, as in China, individual regions have led the national governments in environmental protection efforts due to citizen interest or especially motivated local elites. In Japan, for instance, Kawasaki City approved a "basic environmental plan" that was more stringent than the national government's plan to improve air, water, and soil quality and to enhance energy efficiency in its urban heating. In Minimata, where mercury poisoning in the 1960s affected people in areas around Minamata Bay and resulted in the birth of environmental activism in Japan, the municipal government has now transformed the region into a model environmental area. It established an extensive recycling program in which residents separate their household garbage into twenty-three categories, which is further refined into eighty-four types. In 1999, about five thousand people representing corporations and local governments visited Minimata to learn from the city's experience.[97]

In some of the wealthier, more industrially advanced countries of Asia, businesses are playing an active role in advancing environmental protection. In Japan, for example, Toyota Motor Corporation invited 450 of its suppliers to a meeting at which the company asked them to earn ISO 14001 certification, which connotes that companies have attained international standards of environmental production, within four years. By summer 2000, about 200 companies had earned the certification.[98] Japan's Minebea Group and its electronics and bearings plants have won acclaim not only in Japan but also worldwide for their work on ozone depletion and their efforts to support environmental protection in cities such as Shanghai.[99]

Pushing the Boundaries: The NGO Sector

Throughout the region, however, one of the most potent forces for environmental protection has been NGOs. Along with the media and international NGOs, domestic NGOs have played a crucial role in raising the profile of environmental issues both among government elites and within the general populace. They transmit information, educate, monitor the implementation of laws, and challenge official findings. Often they provide important technical expertise, contributing to both the identification and the solution of a problem.

Once unleashed, governments have found NGOs a difficult force to contain. Frequently they go beyond the mandate of the central government, challenging not only local actors but also central officials. In some cases, they have become enmeshed in broader political movements, providing cover for democracy and human rights activists.

Indonesia and Malaysia evidence the complexity of the relationship between government and NGO activity. Here, the governments have tried to co-opt the NGOs. The Indonesian Ministry of the Environment, for example, in the early 1990s called on NGOs to supplement its weak capacity and help mobilize public support for government programs such as its Clean River Program, which worked to clean up twenty-four of the country's most polluted rivers. The Ministry of the Environment actively encouraged NGOs to ally with community groups and the media to monitor the progress of local industry efforts. By 1993, more than two hundred enterprises had agreed to decrease their pollution loads, and the water quality of the rivers significantly improved.[100]

At the same time, watchdog organizations such as the press and NGOs were carefully monitored to prevent an "all out assault" on vested interests and government policy.[101] In the early 1990s, for example, local villages in Sumatra protested deforestation and water pollution resulting from a rayon factory. The government closed down an NGO and arrested and intimidated the protesters. In response, Wahana Lingkungan Hidup Indonesia (WALHI; the Indonesian Forum for Environment), a national-level forum of NGOs, filed Indonesia's first ever environmental court case challenging both government and industry compliance with environmental regulations. Although the case was lost, it set a precedent and demonstrated a new level of environmental NGO activity.[102] More recently, in 2001, four Indonesian NGOs threatened to call for an international boycott by Indonesia's aid donors if the government did not take action to halt illegal logging.[103]

In Malaysia, in 2001, environmental NGOs originally supported a government and business-led effort to encourage Malaysian businesses to certify voluntarily that their timber products were derived from sustainable resources. This certification process has become increasingly important in international trade in timber products. However, in early 2002, the fourteen NGOs and community-based organizations withdrew their support, claiming that their participation was being inappropriately marketed as signifying NGO approval for the plan.[104]

Such independent, proactive behavior is not typically welcomed by authoritarian governments. As the Malaysian energy minister commented, "Environmentalists are not qualified to make comments on the economic feasibility of such projects." Yet some are willing to voice an alternative perspective. One Malaysian court of appeal judge, Datuk Gopal Sri Ram, called on the judiciary to simplify the regulations for public interest litigation on the environment and to strengthen the authorities responsible for environmental protection and enforcement, noting that the environment was suffering as a result of the lack of public participation in environmental protection.[105]

In the Philippines, Thailand, and Korea, NGOs have extended themselves well beyond the boundaries of environmental issues and have been frequent and vocal opponents of state policy. During periods of military or authoritarian rule in each of these countries, environmental issues were closely entwined with broader political protest movements.

In Thailand during 1974–1975, a student opposition movement protested the government's decision to grant valuable, illegal mining concessions to the Thailand Exploration and Mining Corporation (TEMCO). It was the first time that the authority of the military government had been so challenged. Eventually, the students won, the government rescinded the concessions, and environmental groups blossomed throughout the country.[106] Over time, NGOs have become an accepted part of the Thai decision-making process, aided by several factors. As the Thai middle class grew during the 1980s, their concern over environmental degradation increased, the military government opened the space for popular participation on environmental issues, and the international community stepped in with substantial financial and organizational assistance.[107] Thai NGOs, for example, worked with the Electric Authority to reconfigure the Pak Mun hydropower project to minimize the number of people who would need to be relocated due to flooding. This process was facilitated because the government fully expected that its plans would undergo public review before the project was initiated.[108]

In the Philippines and South Korea, environmentalism became even more closely linked to calls for democracy. Under the rule of authoritarian dictator Ferdinand Marcos during the early 1970s, environmental protests in the Philippines were closely integrated into the broader struggle for democracy. The democratic movement shaped the agenda of the environmental movement by highlighting the link between environmental degradation and authoritarianism and pointing out the abuses that occurred from the "exclusion of public participation in development planning, the concentration of resources in a few hands, and the intolerance to alternative development strategies fostered under undemocratic political and economic structures."[109]

By the time Corazon Aquino assumed power in 1986, NGOs had become a strong, independent force, in large part because international donors were channeling aid through the NGOs rather than through a government that was perceived as corrupt and inefficient.[110] In 1997, the government halted construction of a coal-fired power plant outside Manila, largely due to political pressure by the environmental NGO, Crusade for Sustainable Development. The group proved that the power company and local officials had colluded to conceal public opposition to the project and that there had been numerous environmental violations.[111]

In South Korea, democracy activists often used environmental movements for political cover; throughout the 1970s, environmental protection was one of the few issues over which the military regime permitted public protests.[112] At the same time, the environmental movement learned from the democracy movement, adopting its strategies of protest and confrontation.[113]

As in the Philippines, as South Korea began the transition to a democracy in 1987–1988, the environmental movement shifted its tactics from active confrontation to peaceful action, putting a priority on "creating a mass base, acquiring expertise . . . and [seeking] mass media attention and public sympathy."[114] In 2000, Green Korea United, one of the largest environmental organizations in Korea, filed a lawsuit against the government via one hundred children throughout the country, alleging that the government had permitted the quality of life to deteriorate with its "reckless and unruly urban development."[115] Green Korea United also launched a successful campaign to prevent Taiwan from shipping its nuclear waste to North Korea.[116]

Thus, NGOs have sometimes been a powerful force in setting the environmental agenda in Asian countries. Yet many governments remain wary of NGO activity and limit their activities through various restrictions and requirements. Some of this wariness stems from the link that has emerged between environmental activism and broader issues of social change. According to one editorial in a Malaysian newspaper, this link has two negative consequences. First, it suggests that environmental NGOs are remiss in not focusing on their core issues such as the "illegal infrastructure encroachment in national parks and invasion of watershed areas."[117] Second, and undoubtedly of greater concern, is the belief that environmental NGOs may use their base of public support to agitate for human rights. As the editorial noted,

It is not an exaggeration to say that pure environmental movements today have become intertwined. In fact they could even be described as being a prisoner of the 1990s global spread of concern for human rights. This variation was born out of Europe and has become very effective, particularly in promoting democracy in developing countries. . . . It may, therefore, be time to reconsider the perception of NGO groups so that the entire community is not discredited by human rights groups that disguise themselves as environmental NGOs. In this way the legal and technical matters are not confused by sentiment or ideology that are western-based rather than relevant and home grown.[118]

In wealthier Asian nations, such as Japan and Taiwan, environmental activism has progressed along two distinct paths, much like in the United States. First, are the NGOs concerned primarily with local or "not in my backyard" issues, such as polluting enterprises, incinerators, or nuclear power plants. Such groups might utilize traditional rituals, religion, and folk festivals to agitate for change. Taiwanese activists, for example, organized a parade in opposition to China Petrochemical Plant, including martial arts performing groups with real weapons; their performances were threatening enough to cause one thousand riot police to back down.[119]

A second, sizable element of the environmental movement in Taiwan is devoted to global environmental issues. Because the movement has been advanced largely by intellectuals trained in the United States, its focus has mirrored that of the West, emphasizing global issues such as deforestation, nuclear power, the ozone layer, and climate change.[120]

In Japan, the vast majority of Japanese NGOs are locally based; only 9.5 percent consider themselves national organizations. Their primary focus is nature protection, and secondarily global environmental protection and pollution. They direct the majority of their energy toward recycling, antipollution campaigns, and organic food and other cooperatives.[121]

The political structure in Japan is not conducive to a flourishing NGO community. For example, gaining tax-exempt status requires either capital assets of about $2.5 million or a vast membership. (The number of members needed is determined by the ministry with which the NGO is affiliated.)[122] In addition, NGOs rely heavily on government funds, especially from the Environment Agency, which during the mid-1990s, was criticized for disbursing money to "industry affiliates and quasi-governmental organizations" rather than genuine environmental NGOs.[123] Finally, NGOs have few entry points into the Japanese political structure; there are no provisions, for example, for public participation in environmental policy making or implementation. Environmental litigation is also rare because the scope of activities that may be challenged by environmental activists is narrow, and the highest court tends to support the government.[124]

Despite the decades-long history of environmental NGO activity throughout the region, many NGOs face overwhelming difficulties. In Japan, the Philippines, and Thailand, NGOs rely heavily on external funding: in Thailand, many NGOs receive as much as 80 percent to 90 percent of their financing from overseas supporters.[125] And in several Asian countries, the media is still controlled to a large extent by the government, which limits access to information.

Government and business have also co-opted many Asian NGOs. In Thailand, for example, while grassroots NGOs still protest dam construction, mining, and urbanization, big business has fostered a "cooperative" approach toward environmental protection pursuing green technology and beautification campaigns.[126] They thereby escape public censure and, to a large extent, neutralize potential criticism from environmental activists.

Nonetheless, environmental NGOs have pushed the envelope of environmental protection throughout the Asia Pacific region, challenging central government plans and, in some cases, mobilizing the populace for local environmental protection concerns. While restricted in activity—whether for economic or political reasons—they nonetheless provide one model for China's future environmental protection efforts.

Lessons for China

Throughout the Asia Pacific region, Eastern Europe, and the Soviet Union, during the 1980s, the tale of environmental protection is a relatively straightforward one. Senior government officials placed a premium on economic development, believing that it held the key to social stability. The environmental consequences of this relentless drive to develop, however, inevitably began to damage public health, the standard of living, and society's overall well-being.

The leaders' initial response to their environmental challenges is a political one: to develop an environmental protection apparatus with a central agency (often with regional or local bureaus); to promulgate an array of laws; and to open a degree of political space for citizen involvement. Two decades later, in countries that have prospered economically such as Japan, Korea, Poland, Hungary, and the Czech Republic, government leaders are enabling these environmental bureaucracies, grassroots efforts, and legal infrastructures to fulfill their missions by devoting greater resources to environmental protection and supporting the legitimacy of environmental actors. In the central European states, international pressure from the EU also has been a critical factor in energizing the governments' commitment to environmental protection.

Still, the majority of governments in Asia, central Europe, and Russia remain focused overwhelmingly on economic development to the near exclusion of environmental protection. They avoid the more difficult choices—to invest substantially in environmental protection, to assess heavy fines for polluting enterprises and other environmental malpractice, and to raise the prices for natural resources— fearing that such policies will jeopardize economic growth. Their integration into the international economy is a mixed blessing, bringing both international environmental assistance and access to better environmental technologies from trade and foreign investment but also the potential for more rapid exploitation of their natural resources and import of others' polluting industries and waste. NGOs flourish in some countries, but their efforts often are hampered by legal and illegal means. Even in wealthy democratic nations like Taiwan and Japan, the role of environmental NGOs has been effectively circumscribed by political considerations.

Compared to their counterparts in the United States and Europe, Asian NGOs and media have been less well institutionalized. They often depend on financing from governments and multilateral development banks, considerably weakening their autonomy.[127] In addition, in East Asia, there has not been a "strong participatory culture" or a "well-developed charitable ethic"[128] to support a more proactive environmental movement. In the countries of Eastern Europe and the former republics of the Soviet Union, especially Russia, too, NGOs have encountered significant political roadblocks in attempting to continue their environmental activism. And when governments feel threatened by environmental activism, such as in Russia or Malaysia, for example, they may resort to accusing environmental NGOs and activists of acting against the best interests of the state.

The experience of these countries reinforces much of China's story to date and is suggestive of both future challenges and opportunities in China's environmental protection effort. As we have already seen, economic development has provided the wherewithal for a few regions within China to begin to respond more aggressively to their environmental challenges: investing greater resources, empowering local environmental officials, and attracting environmentally sound foreign investment. The demands for accession to international regimes such as the European Union accelerate a commitment to environmental protection, suggesting that China's accession to the World Trade Organization may, in a more limited fashion, produce similar results.

Still as the majority of countries continue to struggle economically, their environmental institutions remain weak, their willingness to enforce laws and insist on environmentally sound foreign investment remains poor, and their tolerance for wide-ranging fully independent environmental activity remains limited. Much of China seems to be on this environmental trajectory.

Broader Political Change and the Environmental Factor

Beyond the consequences for the environment, however, the path elected by these countries to respond to environmental degradation and pollution has had enormous consequences for the broader political economies of the countries. In Eastern Europe, especially in

Bulgaria, Poland, Hungary, and Czechoslovakia, environmental groups and activists were key players in the downfall of communist regimes.

Green movements throughout Eastern Europe shared several features. They were rooted in the intelligentsia (writers, academics, scientists, and other professionals) and the younger generation, especially college students.[129] They also boasted prominent intellectual leaders who were interested both in political change and environmental reform. For example, Alexander Dubcek—who had led Czechoslovakia during the 1968 Prague Spring, which brought six months of political openness in the country before Soviet troops invaded—was a forester. In Armenia, one of the first environmental protests was led by 350 intellectuals, such as the writer Zori Balayan, who petitioned general secretary Mikhail Gorbachev to address the devastating air pollution problem in Yerevan, the capital.[130] Yury Shcherbak, a famous cardiologist-writer, also served as the head of Zelyonny Svit (Green World), a NGO organization in Ukraine.[131] He later became Ukraine's environmental affairs minister.

Environmental movements also served as a "school of independence for an infant civil society." Glueck has termed this a "shelter function"—these movements or organizations provided cover for grievance sharing and civil association.[132] As Glueck eloquently states,

The environmental appeal drew on deep chords of feeling: a sense of assaulted nationhood, moral pollution, and social helplessness. Environmental degradation occurred under the direction of bureaucracies unaccountable to local communities; it occurred amidst a deepening sense of social despair; and it occurred without freedom of information, without public review, and without the secure right of remonstration. These frustrations, along with the infant associations of civil society, blossomed within the politics of ecology.[133]

He goes on to note,

Green campaigns and organizations were meeting grounds for liberally-minded individuals, breeding grounds for activist citizens and alternative elites, and training grounds in organization and tactics. They were an education in self-initiative for people long reconciled to waiting for state directives and in cooperation for people used to distrusting their neighbors. Utilizing the resource of their being "apolitical," they undertook activities that challenged the State's control over communication and human mobilization.[134]

Glueck also points out the "values function" performed by environmental movements; that is, especially among youth and the intelligentsia, environmentalism provided a new world view that was devoted to a more "harmonious, post-materialist, and caring politics—one directly opposite the consumerist and self-directed values promoted by neo-totalitarian life."[135] Glueck cites the Hungarian campaign against the Danube dam as emblematic of this function. The dam brought together a wide range of value-oriented concerns: *environmentalism* (the dam would destroy an irreplaceable ecosystem, and Hungary's largest underground water source would lose its self-cleansing capacity and eventually dry up); *nationalism* (the dam would mean the destruction of a historic national site); and *democratization* (the dam represented an opportunity to mobilize the public in criticism of the Communist Party).[136]

In some cases, then, transboundary pollution issues or large, unpopular government projects served to raise the flags of both environmentalism and nationalism, thereby forging broad societal consensus against both the project and the ruling government. In otherwise totalitarian states, environmental issues provided a state-approved outlet for public discontent. Environmental degradation and other political considerations fueled awareness and political activism, environmental NGOs became the lightning rods through which popular social and political discontent was channeled. The result was not necessarily an improvement in the state of the environment but, in several cases, the reconfiguration of the entire system of governance.

In the Asia Pacific region, environmental activists also became closely linked with democracy activists and agitated for broader political reform. In these instances, environmental issues did not serve as a catalyst for regime change but did permanently enlarge the political space for social action.

Political Change and the Environment in China

Whether NGOs in China will play a role similar to those of Eastern Europe, the former Soviet Union, and Asia in advancing broader political change remains an open question. In some of the former republics of the Soviet Union and countries of Eastern Europe, forces for political change were galvanized around a single environmental issue, like the Danube dam in Hungary or the Chernobyl nuclear plant in Russia and Ukraine. In addition, disparate religious, human rights, labor, and environmental interests were able to coalesce around political change because they had developed means of communication and a formal organizational structure with clear goals and committed leaders. At least three necessary factors were present in these other countries that enabled NGOs to engage in advancing broader political reform: (1) an ability to tap into existing broader societal discontent; (2) links with other types of NGOs and an ability to communicate within the NGO community; and (3) coalescence around a particular environmental challenge, often involving governmental corruption or lack of transparency.

In China, societal discontent is evident everywhere. It is expressed in forms as diverse as mounting labor unrest, peasant protest, and increased religious activity. As the government has diminished its role in guiding the economy, its role in managing society has decreased as well. For this reason, it retains few levers to shape public opinion and action, with the exception of suppression. It is this discontent, if mobilized throughout the country and more specifically directed at the Communist Party, that Chinese authorities fear.

Chinese leaders are especially wary of the potential for religious organizations to become conduits for broader social discontent. In general, the Chinese government has permitted religion to flourish during the past two decades. However, the reform process is now fostering conditions for the politicization of religious organizations. The current political and economic uncertainty, the public perception of a lack of virtue and integrity among Chinese officials, and the role of religious groups in meeting social welfare needs such as education and medical care, all suggest that religious organizations might become a vehicle through which Chinese society is mobilized for political change. Moreover, China's pre-1949 history is replete with examples of challenges to political authority by religious movements and secret societies. Hence, Beijing's full-scale attack against practitioners of Falun Gong, the spiritual and exercise movement, after ten thousand adherents gathered around the leadership compound Zhongnanhai in Beijing in April 1999.

The widespread discontent in China's countryside is another worry for the Party leaders. In Hunan Province in January 1999, for example, more than ten thousand peasants, organized by a two-year-old illegal organization of local farmers (Volunteers for Publicity of Policies and Regulations) rallied to protest excessive taxes and the corruption of local CCP officials. They were met by hundreds of police and a riot squad from the provincial capital. The leaders of the organization fled to Beijing to try to appeal to the central party leadership, where they were later arrested.[137]

Cities, especially in China's northeastern rustbelt, have also become sites for frequent protests. In March 2002, between thirty to fifty thousand oil workers in Daching protested the giant oil company PetroChina's failure to fulfill its pension obligations to its retired workers. In another city in Liaoning Province, Liaoyang, labor leaders organized tens of thousands of Chinese workers and pensioners in March 2002 to protest low retirement pay. Although the organizers of the protests were quickly arrested—indeed swift reprisals are the rule whenever a protest appears to be well planned and organized—the workers continued to demonstrate, calling for their leaders' release.

Such protests are not likely to abate in the foreseeable future, as Beijing appears increasingly unable to meet the social demands of its citizens.

Based on the experience of other countries, what would it take for China's nascent environmental movement to become a real political force, tapping into this broader societal discontent? Three elements would be essential: (1) a unifying aspiration, (2) a means of communicating, and (3) an issue or event that would serve as a catalyst for action.

First, there would need to be a goal that unifies disparate interests. Some environmental groups, such as the China Development Union (CDU), have tried to articulate such unity. Registered in 1998 by a former journalist and businessman, Peng Ming, in Hong Kong under the China Cultural Exchange Company, the CDU, which boasted seven thousand members, had as its formally stated goal addressing issues such as the environment, unemployment, and corruption. In his 1998 book, *The Fourth Landmark*, Peng argued that "China's industrialization policies over the past half century would cause enormous environmental disasters."[138]

The CDU's mission statement articulates a loose linkage between environmental protection and democracy:

> The purpose of the CDU is to unite all the Chinese people of insight and related organizations and accelerate the transformation of the Chinese nation from the industrial civilization to the ecological civilization. The CDU works on the fulfillment of human being's perpetual living and the healthy development of human nature as necessary for an ultimate realization of the sustainable development of the state and the happiness of the people. Presently, CDU modulates its current point, one of its most important, on pushing forward the political reform, the establishment of a democratic and constitutional government, and the amelioration of human right condition.[139]

As Peng stated, "Our goal is not to destroy the old order but to create a new one." Not surprisingly, the Chinese authorities became alarmed at Peng's intentions. In July 1998, Peng invited Fang Jue, a well-known political reformer who had earlier distributed a public statement calling for democratic reforms based on a Western system of checks and balances, to discuss strategies for political change.[140] In October of that year, Beijing shut down the CDU and detained Peng as he attempted to flee to Hong Kong.[141] He served eighteen months in a labor camp before being released. At the end of 2000, Peng fled to Thailand, and then on to the United States, where he continues to agitate for political change in China.[142] During spring 2003, for example, Peng founded the China Federation Foundation, which seeks the forceful overthrow of China's communist government. His vision for China includes religious freedom, freedom for NGOs, political parties, and rule of law. Yet his approach to overthrowing the government, including encouraging a collapse of the financial system and/or violence, has alienated him from many other Chinese democracy activists based abroad.

The plight of He Qinglian, the acclaimed Chinese journalist and author of *China's Pitfall* (1998), further suggests the precarious position of those who link environmental protection to broader social, political, and economic challenges. While He is not per se an environmentalist, her work includes discussions of environmental degradation in the broader context of the failings of the CCP, such as corruption, lack of accountability, and lack of transparency. During 2000, He lost her job as an editor at the *Shenzhen Legal Daily* for her outspokenness, and in June 2001, fearing imminent arrest, she fled to the United States.

Second, there would need to be a means of communication among NGOS. One possibility is the Internet. The China Democratic Party, for example, has used it to communicate both within and across provincial boundaries without alerting Beijing, demonstrating that this vehicle is a potent force for organizing subversive activity. The SARS (severe acute respiratory syndrome) crisis in spring 2003 also demonstrated the potential of instant messaging as a means of communicating within society.

Finally, there might well have to be a catalytic event or disaster to tap into and ignite the widespread but often latent dissatisfaction that many in China feel concerning the CCP and the current social and economic situation. The Chinese leadership, however, has proved adept in handling such situations (for example, the 1998 flooding of the Yangtze River), rising to the occasion and even enhancing its legitimacy in the process. Without such a catalytic event, a more likely future path for China may well be that taken by some of its Asian neighbors, with environmental NGOs serving as a constant force for reform, exposing corruption, demanding political transparency, and contributing to the gradual evolution of the political system.

Still, linkage between environmental politics and political change in China may arise from yet another source. In January 2001, the Chinese Academy of Social Sciences published a study reviewing the lessons of the collapse of the Soviet Union for China's leaders. The book, *The Collapse of a Superpower: Deep Analysis of the Reasons for the Soviet Disintegration (Chaoji Daguo de Bengkui—Sulian Jieti Yuanyin Tansuo)*, is noteworthy for its conclusion that "political dictatorship, ethnic chauvinism, and diplomatic hegemonism" were the reasons for the USSR's collapse. The authors called on the Beijing leadership to move more rapidly toward political reform to avoid a similar fate.[143] Environmental NGOs, with their uniquely nonthreatening message and form of mass activism, might be tapped to play a role in developing their own political platform.[144]

For China's leaders, the message of history is clear. While involving citizen participation in environmental protection, encouraging regional initiatives, and supporting the efforts of private business may yield some important gains in improving specific environmental problems, these strategies are not sufficient. Moreover, an environmental protection apparatus that is weak and undermined by corruption or by more powerful industrial interests only fans the flames of environmental protest. In extreme cases, once the door has been opened to the formation of social groups, like environmental NGOs, other social interest groups focused directly on advancing political change can be expected to join the cause. Few countries have managed to retain an authoritarian government while permitting nongovernmental action to thrive.

CHAPTER EIGHT

AVERTING THE CRISIS

China's leaders face a daunting task. With one-quarter of the world's population, centuries of grand-scale campaigns to transform the natural environment for man's benefit, intensive and unfettered economic development, and, most recently, its entry into the global economy, China has laid waste to its resources. The results are evident everywhere. Water scarcity is an increasingly prevalent problem. Over one-quarter of China's land is now desert. China has lost twice as much forested land over the centuries as it now possesses. And air quality in many major cities ranks among the worst in the world.

Of equal, if not greater, concern than the immediate environmental costs of China's economic development practices, however, are the mounting social, political, and economic problems that this clash between development and environment has engendered. China's leaders must now contend with growing public health problems. Rising rates of cancer, birth defects, and other pollution-related illnesses have been documented throughout the country. The public health crisis also contributes to growing numbers of protests, some peaceful and some violent, as the government, either through incompetence, corruption, or lack of capacity, proves incapable of taking appropriate action to address the people's concerns.

The economic costs of China's environmental degradation are rising sharply. Most immediately, poor air and water quality has direct costs in terms of crop loss, missed days of work from respiratory diseases, and factory shutdowns from lack of water. Even greater challenges are on the horizon. Several of China's major river systems are running dry in places, necessitating huge and costly river diversion schemes. Much of China's north is under increasing threat of desertification, prompting vast afforestation schemes, with only mixed results.

These depleted land and water resources, coupled with the river diversions, will contribute to migration on the scale of tens of millions over the next decades. While this will relieve population pressure on some of China's most overgrazed and intensively farmed land, it will increase the strain on many urban areas. Already cities such as Shanghai are experiencing significant stress to their sanitation and waste systems, as well as difficulty in gaining access to natural resources such as water.

At the same time as the reforms have exacerbated old (as well as introduced new) environmental challenges, they have not managed to break free of other aspects of China's environmental legacy. Particularly damaging has been Beijing's continued reliance on campaigns to address vast, often complex environmental problems. History has demonstrated repeatedly that the challenges of deforestation, pollution, and scarcity of natural resources are poorly addressed by grand-scale campaigns that attend little to the complex social, economic, and environmental/scientific issues that underpin these challenges. Moreover, even as China assumes a leadership position in the global economy and the international community, its leaders struggle to move beyond traditional notions of security that contribute to large-scale development programs with potentially highly deleterious environmental consequences, such as the grain self-reliance and "Go West" campaigns.

Yet there are signs of hope.

China's post-Mao leaders have developed a far more institutionalized system of governance, with a codified system of laws. This is a critical step forward for environmental protection. In the 1970s and 1980s, the leaders established an environmental protection bureaucracy, held a series of important meetings related to the environment, and issued laws and regulations to strengthen environmental protection throughout the country. Official investment in environmental protection by the central government—practically nonexistent in the 1970s—increased to 1.3 percent of the country's gross domestic product (GDP) at the turn of the century.

Over time, Chinese environmental protection officials have become increasingly adept at developing new policy approaches, drafting laws, and, to a lesser extent, enforcing them. In many instances, they have taken advantage of other aspects of the reform process—utilizing expertise and funding from abroad and relying on growing grassroots and media pressure for better enforcement at the local level.

Still, by most measures, the central environmental protection bureaucracy in China remains weak. With roughly five times the population of the United States, China possesses a central environmental protection bureaucracy only one-twentieth as large. Central government funding for environmental protection, while increasing steadily over the course of the reform period, is still well under the level that Chinese experts claim is necessary to prevent further deterioration. Since 1998, there exists no central environmental agency or commission capable of convening the full range of ministries necessary to resolve many complex environmental challenges that cross bureaucratic boundaries. Even as laws are passed, administrative decrees issued, and regulations set, the politics of resource use conspire to undermine environmental efforts; and the lack of a strong legal infrastructure has enhanced opportunities for corruption and resulted in a systemic crisis for environmental protection enforcement.

Yet, it seems plausible that the small central environmental protection apparatus and its relatively weak reach are at least partly by design. Through the reform process, the central government evidences the belief that, as with the economy, reliance on local authorities and the nonstate sector will yield a better outcome than strong central governance.

Devolution of significant authority for environmental protection to local officials represents one critical element of this process. It has resulted in significant variability in China's environmental situation from one region to the next, with some regions and localities aggressively pursuing stronger environmental protection efforts while others fall further behind.

Cities such as Dalian, Shanghai, and Xiamen routinely invest a significant percentage of their local government revenues in environmental protection; reach out to the international community for extensive technical and financial support for environmental concerns; and have developed relatively well-staffed, well-funded, and well-supported local environmental protection bureaus (EPBs). In all these cases, the driving force was a mayor who perceived his personal advancement and/or the reputation of his city as linked to an improved environment, although economic development clearly provided the wherewithal to address the region's environmental challenges.

As was the case centuries ago, however, only those regions with enlightened local officials and substantial local resources—whether derived domestically or from the international community—have managed to control, if not abate, environmental pollution and degradation. In many other instances, this devolution of authority has contributed to poor efficacy in responding to local environmental challenges. Many local EPBs are grossly understaffed and underfunded and, most important, beholden to local governments for their livelihood. Not surprisingly, they generally lack the political clout within the local bureaucracy to monitor, much less to respond effectively to, local environmental problems. The weak financial and enforcement links between Beijing and provincial and local governments compound the problem: central directives are rarely, if ever, well executed. In poorer regions, such as Sichuan Province, even when the officials recognize the necessity of improving their environmental protection efforts, they are stymied by a lack of resources.

A third aspect of the reform process, China's decision to join the international economy and the international community writ large, has also equipped China's leaders with a vastly different set of political and economic tools than their predecessors possessed. China has opened its arms to embrace technological assistance, policy advice, and financial support from the international community to help improve its environmental situation.

Environmental governance in China increasingly incorporates not only technologies but also norms from abroad. China's participation in the United Nations (UN) Conference on the Human Environment in 1972 and the UN Conference on Environment and Development in 1992 sparked enormous change in the environmental attitudes and orientations of Chinese elites, which have begun to trickle down through education to the wider society. The 2002 UN World Summit on Sustainable Development in Johannesburg may well have a similar impact, already having served as the occasion for China to announce its decision to ratify the Kyoto Protocol to address the challenge of global climate change. The Olympics in 2008 similarly have the potential to drive important advances in environmental technologies, encouraging, for example, the introduction of natural gas into Beijing as a replacement for coal and the purchase of a range of alternative fuel vehicles by the local government. How far the environmental impact of the Olympics will extend beyond Beijing, however, remains to be seen.

Outside actors have also taken advantage of China's economic reforms to introduce new policy approaches to environmental protection. International governmental organizations, such as the World Bank and Asian Development Bank, individual countries such as Japan, and international nongovernmental organizations (INGOs) have been instrumental in persuading their Chinese colleagues to experiment with market-based mechanisms such as price reform for natural resources, tradable permits, and energy service companies. In some cases, such as pricing reform for water, resource scarcity and population pressure provided the opening for the reform to be initiated, giving credence to the more optimistic view of how population affects the environment.

Further, China's accession to the World Trade Organization (WTO) will place new strictures on its some of its production practices, particularly in agriculture. Several Chinese products have already been deemed insufficiently green, resulting in a sharp drop in those exports to WTO member countries. Yet, the WTO also will offer some highly polluting industries such as textiles the opportunity to ratchet up significantly their exports.

And there are concerns over the implications of China's burgeoning levels of foreign trade and investment for the environment more broadly. Many multinationals bring only their best environmental practices and most advanced technologies to China. Royal Dutch Shell's broad engagement in China's environmental protection efforts—funding environmental NGOs, insisting on top-quality environmental and socioeconomic impact assessments for its pipeline venture, and bringing natural gas to the coastal provinces—offers a good example of the ways in which international firms can shape China's environmental trajectory in a constructive manner. Many others, however, offload their most polluting enterprises on regions desperate for foreign investment.

China's weak enforcement of its own environmental protection laws also undermines the potential environmental advantages of foreign direct investment. Many multinationals complain that despite their best efforts, local officials and enterprise managers prefer not to use the pollution control technologies they provide in order to decrease the costs of operating the plants. Or, in other instances, foreign firms simply cannot compete against domestic firms that do not abide by the country's environmental regulations.

China's leaders have also placed the future of environmental protection in the hands of the Chinese people, opening the door to grassroots activities, NGOs, and the media. This has produced an exciting and vibrant—if still nascent—source of policy approaches and enforcement capabilities. Environmental NGOs, independent environmental lawyers, and an aggressive media all have injected new ideas and methods into China's environmental protection efforts, reflecting a gradual evolution in the traditional belief of the Chinese people that it is the role of the government to safeguard the environment. Moreover, NGOs and the media have begun to reshape the environmental and political consciousness of the public and some government officials in ways that are important not only for environmental protection but also potentially for broader social and political purposes.

Yet, even as the NGO sector and media have become highly valued adjuncts to China's environmental protection work, restrictions governing their work still remain. The Chinese government is wary of the potential for NGOs and the media to move beyond issues of local enforcement to criticize central government policy or potentially serve as a force for broader social change. Thus, in many respects, China's environmental NGOs have developed far more slowly over the past decade—both in terms of the absolute number of such NGOs, their membership, and the scope of their work—than might have been anticipated.

Finally, this is a circular process: Even as political and economic reform shape China's environment, the environment, too, is influencing China's reform process. Environmental pollution and degradation are costly to Chinese economic productivity, contribute to tens of millions of environmental migrants, damage public health, and engender social unrest. In addition, the environment may serve as a locus for broader political discontent and calls for political reform, as it has in other countries. Thus, China's reform process has brought an extraordinary dynamism and energy to both the nature of China's environmental challenges and its environmental protection efforts; yet it also increases the uncertainty in attempting to chart China's future environmental path.

Scenarios for the Future

Given the dynamic nature of both China's economy and its evolving political system, assessing China's environmental future and its broader implications for the country and the world is no easy task. Integrating centuries of historical attitudes and approaches with the current infusion of new ideas and technologies into this interplay of economic development and environmental governance demands consideration of a range of potential outcomes.

Using China's economy as a starting point, three scenarios suggest themselves for China's environmental future, with significantly different implications for both the country and the world.

China Goes Green

In the first scenario, China's economy continues to grow, producing more challenges for the environment but simultaneously spurring greater investment in environmental protection at both the local and national levels. Not only the economy and the environment benefit but China's political system is also enhanced through more effective application of the rule of law, greater citizen participation in the political process, and the strengthening of civil society.

In this scenario, China's most vibrant cities such as Shanghai and Dalian serve as genuine models for other coastal and inland cities interested in attracting greater foreign investment and recognition for their livability. As urbanization continues, satellite cities replicate the better environmental practices of the major urban centers rather than becoming dumping grounds for the cities' most polluting enterprises. Shanghai becomes a center for the most advanced environmental thinking and cleaner production, prompting a booming industry in environmental technologies. Beautification campaigns clean up the city streets, and Shanghai develops a high-speed transit system to its satellite cities, thereby sharply slowing the growth rate in car use in the city and surrounding environs. The continued increasing wealth of the city and environmental education opportunities in schools spawn a highly energized green movement, which promotes recycling, water-saving measures, and other grassroots efforts to protect the environment and the city's natural resources. In ten years, the water in the Huangpu River is once again safe for recreation and drinking, leading to a lively riverfront community.

At the same time, Premier Wen Jiabao uses his environmental expertise to push for broader environmental change, and the political success of former Dalian mayor Bo Xilai encourages other mayors to follow suit, using the environment as a stepping stone to positions of greater political prominence. Throughout the country, tens of thousands of model environmental cities sprout, providing China's citizens with unprecedented access to a better future for themselves and their children.

China's entry into the WTO reinforces positive trends in the quality of goods produced, the development of a legal infrastructure, and stronger enforcement capability. Electricity supplied from the Three Gorges Dam and natural gas provided by the West-East pipeline dramatically improve China's energy mix; coal use decreases as the desire for efficiency and higher quality of life becomes paramount. Environment-related public health concerns diminish as China is forced to rethink its rural development and agricultural strategies to accommodate WTO-mandated levels of food safety. As China moves away from intensive farming and toward other, more environmentally sustainable and lucrative crops, the country also becomes world-renowned for the quality and quantity of its organic produce, particularly fruits and vegetables. In the automobile sector, Chinese joint ventures become leaders in producing fuel-efficient cars. Alternative fuel-cell vehicles dominate the roads, encouraged by the desire of Beijing to match consumer interests with the country's diminishing oil supply.

On the governance front, China's NGOs continue to flourish, supported not only by the international community but also increasingly by Chinese citizens who value a clean environment and are willing to contribute both financially and personally to ensure its sustainability. China's top entrepreneurs become an important new source of funding for environmental NGOs, helped by new tax laws and a growing base of wealthy Chinese citizens. Small mass-based environmental NGOs emerge, and the well-established NGOs expand their membership and their mission as the next generation of environmental NGO leaders, led by Wen Bo, Hu Kanping, Li Li, and others, increasingly incorporates international practices of lobbying and lawsuits to protect the environment.

For the Chinese leadership, the environment may change from a political liability into a source of political strength as the people are able to realize both economic prosperity and improved quality of

life. Certainly, environmental issues become increasingly important in local elections as people's concerns broaden past basic living
conditions. Rooting out violations of environmental laws and corruption becomes a source of political credibility for aspiring leaders
and community spirit for mobilized citizens. A formal and organized environmental movement evolves across provincial boundaries
assisted by the media and the growing political strength of environmental NGOs. They advocate a political platform that encompasses
a broad range of social and political interests tied to the environment, including population planning, environmental protection,
rule of law, and economic development ideals. This scenario lends
itself to a deepening or broadening of political reform, as Chinese
citizens increasingly demand that their voices be heard on the full
range of social, political, and economic issues. At the same time,
the increasing confidence of the Chinese Communist Party (CCP)
leads it, on its own initiative, to open the door to greater political di-
versity and political reform of its own accord. By the 2013 presiden-
tial elections, the CCP is recast as the Chinese Socialist Democratic
Union and is one of two or three parties competing for political
power.

The international community—business, government, and NGOs—
joins forces with domestic NGOs and environmentally proactive
government leaders to increase dramatically China's technological
and policy capacity to protect the environment. This partnership
also transforms the Go West campaign into a vast experiment in
sustainable development. Urban planners, conservationists, and
business leaders join forces to develop western China in a model environmental fashion, implementing cleaner production, ensuring
public hearings for new development projects, and establishing local
best practices with solar-powered office buildings, recycling centers,
and state-of-the-art public transportation. The media, together with
the public, serves as a watchdog to ensure that the west does not become a new center for resource exploitation and polluting industry.

For the international community, this scenario offers the potential for improved implementation of international environmental
agreements, declining or stable levels of China's greenhouse gas

emissions, and improvement in China's contribution to transboundary air problems such as acid rain. More broadly, it suggests a China that is more likely to be both willing and able to participate fully in international environmental, as well as other, agreements and institutions. Environmental advocates both within and outside the Chinese government will become important members of the international environmental movement supporting China's participation in the full range of regional and global environmental regimes. And they will use international regimes to hold the central government accountable for its environmental commitments.

Inertia Sets In

This second scenario extends today's status quo. The Chinese economy continues to grow, but greater economic wealth translates only sporadically into enhanced environmental protection. The environment continues to be a drag on the Chinese economy, and both at home and abroad people complain about the worsening condition of China's air and water quality.

In this scenario, the growing use of automobiles, for example, is not matched by more stringent enforcement of the air pollution control law. Automobile use increases dramatically as expected, and low-cost, low-end cars that do not employ the most advanced technologies are favored. Rather than embracing the fuel-efficient, compact cars that populate Europe and Japan, wealthier Chinese follow in the footsteps of Americans, desiring gas-guzzling luxury cars and sport utility vehicles (SUVs). Public transportation becomes a least-desired mode of transport. Despite the spur of the 2008 Olympics, alternative fuel-cell vehicles do not spread in popularity. Air quality continues to suffer, as Chinese officials miss their targets and the number of cars continues to increase.

Top authorities rely on their traditional ineffectual use of campaigns to address nationwide problems, and the Go West campaign confirms the worst fears of local environmentalists: It leads to greater devastation of the natural resources of the interior provinces in the name of development. Political tensions in Xinjiang and Tibet heighten as the two strata of society—the wealthy Han settlers and the indigenous poorer minorities—diverge further.

The devolution of authority to local levels continues to produce a patchwork of environmental protection practices, in which only the wealthiest cities with environmentally inclined mayors use their growing economic capacity to fund improvements in their local environment. There is little transfer of environmental know-how—either technological or policy—to other regions of the country. The northeast continues to suffer the environmental impacts of declining rust-belt industries, and the interior provinces are transformed into a haven for environmentally exploitative practices to lure both domestic and international investment. The costs of remediation projects, massive river diversions, and afforestation campaigns continue to increase as China's leaders continue to favor cleanup rather than prevention. These campaigns serve primarily as public works projects to stimulate the economy and provide work to otherwise unemployed farmers and loggers, but simultaneously drain the already overdrawn coffers of the state banks.

The WTO produces improved enforcement capacity in areas such as intellectual property protection, but this enhanced capability remains narrowly trained on business and does not translate readily into other arenas. Transfer of environmental technologies continues to be hampered by weak incentives and enforcement. Only the World Bank and other international development programs support progress on the environment; China remains unable to develop and sustain a domestically driven environmental technology industry.

With little reinforcement from the central government, NGOs remain constrained in number and in the range of their activities, focusing primarily on species protection and some urban renewal work. They only slowly expand the horizons of their work to lobby against environmentally unsound projects. The next generation of environmental activists becomes frustrated by the political limitations of their work and instead turns more to media and other outlets to express their environmental and political interests. Interest in environmental law develops only slowly because it is viewed as an area with poor future growth prospects.

International actors remain engaged in environmental protection work in China, but are stymied by recalcitrant local actors, weak incentives for adopting new approaches or technologies, and few opportunities for replication of their work beyond single demonstration projects. Foreign investment in environmental industries slows.

For the international community, China remains a partner in international environmental accords, but not necessarily a reliable one. Driven overwhelmingly by their continued desire to grow economically and maintain political stability, China's leaders continue to resist what they perceive as onerous environmental commitments to address problems such as climate change. At the same time, China's contribution to climate change and other transboundary air pollution problems increases as the country continues to develop with small enterprises fueled primarily by coal and as automobiles proliferate.

Environmental Meltdown

There is no guarantee, of course, that China's economy will continue to grow at the 7 percent that some analysts predict necessary to maintain social stability. If indeed there were an economic slowdown, there might be some environmental benefits. Declining industrial production, for example, could lead to decreasing emissions of greenhouse gases. Overall, however, the environment would likely be an early victim.

In the third scenario, China's economy slows or even experiences a significant downturn. Local officials continue to favor economic development at the expense of the environment in an effort to preserve social stability. As a result, China's air quality does not improve, as the country continues to rely on older, more inefficient polluting technologies and automobiles. Water pollution increases throughout the major river systems. And, most important, investment in waste treatment or new conservation efforts diminishes as a short-term outlook prevails.

Especially in the already economically hard-hit areas of northeast China and the interior provinces, massive layoffs in the state-owned enterprise sector and growing problems with environmentally and economically induced migration would also challenge the ability of local governments to provide work for the people. The social welfare system is overwhelmed as continued corruption drains local coffers and impairs the development of a functioning pension system. There are frequent demonstrations, which often turn violent. The outlook for improved quality of life is bleak. Positive environmental trends in forestry and agricultural practices are reversed as logging bans are ignored and farmers attempt to eke out a living on increasingly degraded land.

In the Go West effort, environmental and economic exploitation prompt not only growing political disaffection but also increased violence and protests against domestic and international businesses perceived to support such policies. International business stops investing and withdraws from the interior, which is now viewed as politically unstable. The West-East pipeline becomes a target of sabotage. Weaknesses in the banking infrastructure and law enforcement deepen as Chinese officials seek any means to keep local industry afloat. After making small gains in the rule of law, courts are increasingly reluctant to press enterprises to adhere to environmental protection laws for fear of promoting further social unrest. Corruption continues to erode economic institutions and leadership credibility as Chinese citizens believe they have no recourse for justice.

Unwilling to risk massive layoffs, Chinese officials backtrack on WTO commitments in an effort to preserve Chinese industries. Bureaucracies are given greater latitude to develop regulations that impede the access of foreign companies to China's market. Foreign investors respond by denouncing Chinese practices, further contributing to international friction and trade disruptions. Fewer environmental benefits are also felt due to a slowdown in the influx of foreign technology and increasing acceptance of poorer environmental practices. Gains for China's industry come largely in the highly polluting textile, toy, and tin mining industries; few benefits are reaped in the expected area of agriculture, as China resists opening its market to large-scale grain imports promised in its WTO accession agreement.

In the political realm, this kind of economic downturn could produce at least three different scenarios. The first and most predictable would be a reversal in the present trend of modest political reform, exacerbating the tension within the Chinese leadership between its relatively newfound belief that there is much to gain from embracing forces of local political action and integration with the international community and its traditional fear that such processes will undermine the security of the state and their capacity to govern. In such a time of domestic stress and leadership vulnerability, the environment is not likely to receive much positive attention. More likely, the public security apparatus would increasingly scrutinize the work of NGOs and independent lawyers for the political content of their work. Peasant and worker unrest would be managed through repression and, perhaps, through the projection of an external threat or a manufactured crisis in the Taiwan Straits to rally nationalist sentiment and deflect attention from the country's economic woes.

Although the CCP has proved remarkably adept at harnessing nationalistic sentiment for its own legitimizing purposes, the wave of anti-Party sentiment expressed during the confrontation with the United States over the EP3 spy plane during spring 2001 (when a Chinese fighter jet crashed and a U.S. spy plane was forced to land on Hainan Island) suggests that such views could spiral out of the government's control, producing a new challenge to the stability of the regime.

A second possible outcome would be for China's leaders to pursue a path akin to that of former Soviet General Secretary Gorbachev in the mid-to-late 1980s. Gorbachev responded to economic malaise and his inability to reform the Soviet economy by opening up the political system as a means of diffusing popular discontent and channeling it into institutions such as political parties. This is a less obvious option. Many reform-oriented Chinese leaders have explicitly rejected such an approach. But recent scholarly studies revisiting the lessons of the Soviet experience, as well as visits by senior Chinese officials to Europe to study social democratic parties suggest that discussion has not been closed completely. Reform-minded leaders might well ally with labor, environment, and women's NGOs to explore alternative formulations for restructuring the government to involve greater public participation.

In this scenario, China's environmental NGOs would be a logical locus of political action, drawing on both the intellectual and political skills of the NGO leadership as well as the innate attractiveness of their message and mission for many Chinese. The role played in effective environmental protection by transparency, rule of law, and public participation suggest a natural synergy between movement toward greater political openness and the work of the environmental community in China. Potential alternatives to the CCP might include a green party or a broader "democracy" party that would embrace many of the social issues of greatest concern—labor, the environment, and education among them—that CCP-led development has brought to the Chinese people.

Third is the scenario that many China watchers have previously described: that China might collapse or be riven by wide-scale civil strife. Regime-threatening protests emanating from China's minorities, especially the Uighurs in Xinjiang and Tibetan independence advocates, or from a combination of striking laborers, peasants, and urban intellectuals, are all within the realm of possibility, especially as China's leaders struggle to enforce WTO strictures. In this scenario, the environment could be one of many causes espoused by the protesters. An environmental disaster of significant magnitude, like a collapse of the Three Gorges Dam, could easily serve as the trigger for such protests—a giant symbol of the corruption, lack of transparency, and limited political participation of China's current system.

Which Way? The Role of the United States

Millions of people concerned with the environment, both in China and elsewhere, are hoping that the first scenario of growing prosperity, democracy, and environmental protection becomes reality. But this is by no means assured. Both China's domestic reforms and its deeper integration into the international system require a fundamental change in values. The nationalistic and occasionally violent demonstrations in the wake of the U.S. accidental bombing of the Chinese embassy in Belgrade and the EP3 collision demonstrated that within at least some sectors of Chinese society and the Chinese leadership, the commitment to this kind of change is a tenuous one.

The world community has an opportunity to assist in the process of developing China's approach to environmental protection. The impact of the international community on China's environmental practices has already been substantial. In every regard—technology transfer, governance, and policy development—the international community has helped to shape the evolution of China's environmental protection effort. Fortunately, this involvement shows no evidence of abating and, indeed, seems likely to grow in the future.

Yet much more remains to be done. Technology transfer and adoption of new policy approaches await the development of a stronger legal and enforcement apparatus. Here, the international community, in particular the United States with its strong environmental enforcement apparatus and history of public participation in environmental protection, could be far more active in contributing to the development of China's environmental future.[1]

The environment provides a natural and nonthreatening vehicle to advance U.S. interests not only in China's environmental protection efforts but also in its basic human rights practices and trade opportunities.[2] Several simple steps could be taken to help shape China's future environmental, political, and economic development. Chief among these is removing restrictions on the Overseas Private Investment Corporation and the U.S.–Asia Environmental Partnership. These restrictions, which date to the 1989 Tiananmen Square protests and before, severely limit U.S. influence in China. The United States Agency for International Development (USAID), with its broad emphasis on governance, public health, rule of law, and poverty alleviation could be especially valuable in addressing China's most pressing needs and the United States' most direct interests. For USAID to become actively and directly involved in such activities, however, would require the United States to remove either the prohibition on USAID funding of communist states or the label of communist state from China.

The United States has the chance not only to benefit more significantly from China's current reform process but also to aid in the future evolution of that process in ways that serve broader U.S. political, economic, and environmental interests. This will require the U.S. leadership to develop both an understanding of the opportunities to influence China's future path and a newfound commitment to take the bold steps necessary to do so. It is a challenge that the United States cannot afford to ignore.

Deep U.S. engagement involving both trade and substantial support for political development has been an essential component in the development of our close relations with many nations in Asia, including Japan, Taiwan, and South Korea. As the United States considers the importance of mainland China to regional stability, to the world economy, and to the global environment, it need look no further than its past successes in the region to understand that it has a substantial opportunity at a critical juncture in China's development to play a similarly catalytic role. No matter how extensive China's interaction with the United States and the rest of the world, China itself bears the brunt of the challenge. A future in which China fully embraces environmental protection will require new approaches to integrating economic development with environmental protection. Equally, if not more, important will be a commitment by China's leaders to develop the political institutions necessary to ensure such a future, to bring true transparency and accountability to the system of environmental protection. Speculation remains as to the boldness of President Hu Jintao and Premier Wen Jiabao, China's Fourth Generation leaders. Yet change will require nothing short of boldness. This is a system thousands of years in the making, one in which the greatness of leaders has often been achieved at the expense of nature, through grand-scale development campaigns to control and exploit the environment for man's benefit. But today, greatness may well depend on resisting this tradition and instead developing a new relationship between man and nature.

NOTES

Chapter 1. The Death of the Huai River

1. "Huaihe River Avoids Pollution Disaster—SEPA," Xinhua News Agency, August 7, 2001.
2. Michael Ma, "River Flooding Sets off Pollution Alarm," *South China Morning Post,* August 4, 2001, p. 8.
3. Jiangsu is especially well off, ranking fifth out of thirty-one provinces and autonomous regions. Economic Research Service, U.S. Department of Agriculture, "China Basic Statistics, by Province, 2002," www.ers.usda.gov/Briefing/China/tbl-provinces.xls.
4. *2002 Nian Zhongguo Huanjing Zhuangkuang Gongbao* [2002 Report on the State of the Environment in China] (Beijing: State Environmental Protection Administration, 2003).
5. Jasper Becker, "River Gives Way—to Pollution," *South China Morning Post,* January 13, 2001, p. 10.
6. Hugo Restall, "Examining Asia: A Top-Down Coverup," *Asian Wall Street Journal,* March 14, 2001, p. 6.
7. "China Calls for Rational Exploitation of Major River," Xinhua News Agency, April 19, 2001.
8. Jiusheng Zhang, "Water Resource Protection and Water Pollution Control for the Huai River Basin," World Bank Huai River Basin Pollution Control Project Report, pp. 2–3, www.wb.home.by/nipr/china/pids/huai.htm.

9. Becker, "River Gives Way," p. 10.

10. Mark Hertsgaard, "Our Real China Problem," *The Atlantic Monthly*, November 1997, www.theatlantic.com/issues/97nov/china.htm.

11. Abigail Jahiel, "The Organization of Environmental Protection in China," *China Quarterly* 156 (December 1998): 781.

12. *China Environment News*, July 1993, p. 4.

13. Hertsgaard, "Our Real China Problem."

14. Patrick Tyler, "A Tide of Pollution Threatens China's Prosperity," *New York Times*, September 25, 1994, p. A3.

15. Hertsgaard, "Our Real China Problem."

16. Ibid.

17. Notes provided by Chinese scientist to author, June 1996.

18. "China's Huai River Cleanup Lagging," Kyodo News Service Japan Economic Newswire, November 21, 1997.

19. Joseph Kahn, "China's Greens Win Rare Battle on River," *Wall Street Journal*, August 2, 1996, p. A8.

20. "China Takes Action to Huaihe Polluters," Xinhua News Agency, January 1, 1998.

21. Jasper Becker, "Cracking Down on the Chemical Waste Outlaws," *South China Morning Post*, May 21, 1999, p. 15.

22. *Newsprobe* reporter and producer, interview by author, Beijing, April 2000.

23. Mary Kay Magistad, Robert Siegel, and Noah Adams, "Cleaning Up the Huai," *All Things Considered*, National Public Radio, April 7, 1998.

24. Ibid.

25. "Families Seek Truth over Deaths Near Notorious Polluted Overflow," *South China Morning Post*, June 16, 2000, p. 7.

26. Ibid.

27. "Exploitation Exhausts the Dwindling Huai," *South China Morning Post*, January 14, 2001, p. 6.

28. "Central China Hit by Severe Drought," Associated Press Newswire, August 26, 1999.

29. "Exploitation Exhausts the Dwindling Huai," p. 6.

30. Ibid.

31. Jasper Becker, "Steps for Cleaning up Act on Pollution Announced," *South China Morning Post*, January 1, 2001, p. 4.

32. Jasper Becker, "Clean up of River a Sham," *South China Morning Post*, February 26, 2001, p. 8.

33. "Intelligence," *Far Eastern Economic Review*, March 1, 2001, p. 10.

34. "Firms Ordered to Curb, Stop Operation to Prevent Pollution Disaster," fpeng.peopledaily.com.cn/200108/02/eng200110802_76373.html.

35. "China: Huai River Pollution Control Project," The World Bank Group, March 22, 2001, www.worldbank.org/sprojects/Project.asp?pid=PO47345.

36. "Weekly Monitoring Report on the Water Quality of Huai River Basin," China National Environmental Monitoring Center, February 10, 2003, www.zhb.gov.cn/enlish/SD/river.php3?name=Huai%20River&no=6.

37. Becker, "River Gives Way," p. 10.

38. Karl Polanyi, *The Great Transformation* (New York: Rinehart, 1944), 184.

39. See, for example, Vaclav Smil, *China's Environmental Crisis* (Armonk, N.Y.: M. E. Sharpe, 1993); Abigail Jahiel, "The Environmental Impact of China's Entry into the WTO," draft article, 2002.

40. Yok-shiu F. Lee and Alvin Y. So, *Asia's Environmental Movements* (Armonk, N.Y.: M. E. Sharpe, 1999), 4.

41. Asayehgn Desta, *Environmentally Sustainable Economic Development* (Westport, Conn.: Praeger, 1994), 122–23.

42. Jose I. dos R. Furtado and Tamara Belt, eds., *Economic Development and Environmental Sustainability: Policies and Principles for a Durable Equilibrium* (Washington, D.C.: World Bank, 2000), 75.

43. Ronald Inglehart, "Globalization and Postmodern Values," *The Washington Quarterly*, 23, no. 1 (2000): 219.

44. Jagdish Bhagwati, "The Case for Free Trade," in John J. Audley, *Green Politics and Global Trade: NAFTA and the Future of Environmental Politics* (Washington, D.C.: Georgetown University Press, 1997), 34.

45. Vaclav Smil, *China's Environmental Crisis* (Armonk, N.Y.: M. E. Sharpe, 1993), 192–93.

46. Furtado and Belt, eds. *Economic Development and Environmental Sustainability*, 75.

47. Jahiel, "Environmental Impact," p. 5.

48. Lee and So, *Asia's Environmental Movements*, 4.

49. *Environment and Trade: A Handbook* (Winnipeg, Manitoba: United Nations Environment Programme and International Institute for Sustainable Development, 2000), 3–4.

50. Furtado and Belt, *Economic Development and Environmental Sustainability*, 75–80.

51. Audley, *Green Politics and Global Trade*, 34.

52. John Carey, "Will Saving People Save Our Planet?" in Desta, *Environmentally Sustainable Economic Development*, 85.

53. Desta, *Environmentally Sustainable Economic Development*, 84.

54. See, for example, Geping Qu and Jinchang Li, *Population and the Environment in China* (Boulder, Colo.: Lynne Rienner Publishers, 1994); Angang Hu and Ping Zou, *China's Population Development* (Beijing: China's Science and Technology Press, 1991).

55. Hu and Zou, *China's Population Development*, 191.

56. E. Boserup, *The Conditions of Agricultural Growth* (London: Allen and Unwin, 1965).

57. Carey, "Will Saving People Save Our Planet?" 85.

58. Desta, *Environmentally Sustainable Economic Development*, 80.

59. See David W. Pearce and Jeremy J. Warford, *World Without End: Economics, Environment, and Sustainable Development* (New York: Oxford University Press, 1993), 149; *Environment and Trade: A Handbook* (Winnipeg, Manitoba: United Nations Environment Programme and International Institute for Sustainable Development, 2000).

60. Only in the arena of population policy did the government exert significantly greater rather than lesser control, in an effort to reverse the population explosion that had occurred during Mao's tenure.

61. Hua Wang, Jun Bi, David Wheeler, Jinnan Wang, Dong Cao, Genfa Lu, Yuan Wang, "Environmental Performance Rating and Disclosure: China's Green-Watch Program," World Bank Policy Research Working Paper 2889 (September 2002), p. 4, www.econ.worldbank.org/files/18746_wps2889.pdf.

62. Adam Segal and Eric Thun, "Thinking Globally, Acting Locally: Local Governments, Industrial Sectors, and Development in China," *Politics and Society* 29, no. 4 (2001): 557–88.

Chapter 2. A Legacy of Exploitation

1. See, for example, Richard Baum, "Science and Culture in Contemporary China: The Roots of Retarded Modernization," *Asian Survey* 22, no. 12 (1982): 1166–86; Joseph Needham, *Science and Civilisation in China*, vol. 2 (London: Cambridge University Press, 1956).

2. Baum, "Science and Culture," 1168–76.

3. Derek Wall, *Green History* (London: Routledge, 1994), 2.

4. Samuel P. Hays, *Beauty, Health, and Permanence* (Cambridge: Cambridge University Press, 1987), 13–22.

5. Ibid.

6. Yushi Mao, "Evolution of Environmental Ethics," in *Ethics and Environmental Policy: Theory Meets Practice*, ed. Frederick Ferre and Peter Hartell (Athens: University of Georgia Press, 1994), 45.

7. Ibid.

8. Xiaoshan Yang, "Idealizing Wilderness in Medieval Chinese Poetry," in *Landscapes and Communities on the Pacific Rim*, ed. Karen K. Gaul and Jackie Hiltz (Armonk, N.Y.: M. E. Sharpe, 2000), 104.

9. Rhoads Murphey, "Man and Nature in China," *Modern Asian Studies* 1, no. 4 (1967): 316.

10. Vaclav Smil, *The Bad Earth* (Armonk, N.Y.: M. E. Sharpe, 1984), 6.

11. Yang, "Idealizing Wilderness in Medieval Chinese Poetry," 92.

12. Wenhui Hou, "Reflections on Chinese Traditional Ideas of Nature," *Environmental History* 8, no. 4 (1997): 482–93.

13. Among historical records, different dates are frequently provided for the beginning and end of dynasties. In this book, the dates are taken from William Theodore de Bary, *Sources of Chinese Tradition*, vol. 1 (New York: Columbia University Press, 1960). This book will use C.E. ("of the common era") and B.C.E. ("before the common era") for abbreviation of eras.

14. Geping Qu and Jinchang Li, *Population and the Environment in China* (Boulder, Colo.: Lynne Rienner Publishers, 1994), 16.

15. Mao, "Evolution of Environmental Ethics," 43.

16. Ibid.

17. Carol Stepanchuk and Charles Wong, *Mooncakes and Hungry Ghosts* (San Francisco: China Books and Periodicals, 1991), 53.

18. Wenhui Hou, "The Environmental Crisis in China and the Case for Environmental History Studies," *Environmental History Review* 14, no. 1–2 (1990): 152–53.

19. Charles O. Hucker, *China's Imperial Past* (Stanford: Stanford University Press, 1975), 69.

20. Ibid., 55–56.

21. Richard Edmonds, *Patterns of China's Lost Harmony: A Survey of the Country's Environmental Degradation and Protection* (London: Routledge, 1994), 24–25, quoted in Hong Fang, "Chinese Perceptions of the Environment" (master's thesis, University of Oregon, June 1997), 44.

22. Fang, "Chinese Perceptions," 66.

23. The Tao is "the process by which the cosmos operates. It functions through the interaction of two opposed but complementary and inseparable forces, Yang and Yin, which are perhaps most usefully compared to the positive and negative poles of an electrical system. Yang is associated with the sun, light, and warmth; Yin with the

moon, dark, and cold. . . . When the natural alternation of Yang and Yin is aborted, the result is inappropriate and might be harmful or otherwise disadvantageous; but no question of good or bad arises in any absolute, abstract sense." Hucker, *China's Imperial Past*, 70–71.

24. Ibid., 84.
25. Kenneth E. Wilkening, "A Framework for Analyzing Culture-Environment Security Linkages, and Its Application to Confucianism in Northeast Asia," draft paper prepared for the University of Washington and Pacific Northwest Conference Series on Environmental Security, conference on "Cultural Attitudes and the Environment and Ecology, and Their Connection to Regional Political Stability," University of Washington, Seattle, January 16–17, 1998, www.nautilus.org/esena.
26. Hucker, *China's Imperial Past*, 79.
27. Deshu Xu, *Construction of Safety Culture in China* (Chengdu: Sichuan Science and Technology Press, 1994), 255, quoted in Fang, "Chinese Perceptions," 63.
28. Xu, *Construction of Safety Culture*, 255, quoted in Fang, "Chinese Perceptions," 65.
29. Ibid.
30. Hou, "Environmental Crisis in China," 153.
31. Xu, *Construction of Safety Culture*, 255, quoted in Fang, "Chinese Perceptions," 65.
32. De Bary, *Sources of Chinese Tradition*, 207.
33. Ibid.
34. *Tian*, again signifying the sky, God, and nature, and *zi* meaning the son of *tian*.
35. De Bary, *Sources of Chinese Tradition*, 207.
36. Fang, "Chinese Perceptions," 67.
37. Mao, "Evolution in Environmental Ethics," 43.
38. Helen Dunstan, "Official Thinking on Environmental Issues and the State's Environmental Roles in Eighteenth-Century China," in *Sediments of Time: Environment and Society in Chinese History*, ed. Mark Elvin and Ts'ui-jung Liu (Cambridge: Cambridge University Press, 1998), 587.
39. Ibid.
40. Murphey, "Man and Nature," 314.
41. Hucker, *China's Imperial Past*, 90.
42. Ibid.
43. Mao, "Evolution of Environmental Ethics," 46.
44. Hucker, *China's Imperial Past*, 91.

45. Ray Huang, *China: A Macro History* (Armonk, N.Y.: M. E. Sharpe, 1997), 20–21.

46. Mao, "Evolution of Environmental Ethics," 46.

47. Weiming Tu, "The Continuity of Being: Chinese Visions of Nature," in *Confucianism and Ecology: The Interrelation of Heaven, Earth, and Humans,* ed. Mary Evelyn Tucker and John Berthrong (Cambridge: Harvard University Press: 1998), 113.

48. Ibid.

49. Robert P. Weller and Peter K. Bol, "From Heaven-and-Earth to Nature: Chinese Concepts of the Environment and Their Influence on Policy Implementation," in *Confucianism and Ecology,* ed. Tucker and Berthrong, 322.

50. Ibid., 323.

51. Ibid.

52. Hucker, *China's Imperial Past,* 93.

53. Ibid., 94–95.

54. Ibid., 164.

55. Mao, "Evolution of Environmental Ethics," 45.

56. Ibid., 45–46.

57. Chang-Qun Duan, Xue-chun Gan, Jeanny Wang, and Paul K. Chien, "Relocation of Civilization Centers in Ancient China: Environmental Factors," *Ambio* 27, no. 7 (1998): 575.

58. Mark Elvin, "The Environmental Legacy of Imperial China," *China Quarterly* 156 (1998): 738–39.

59. De Bary does not provide any dates for the Xia dynasty; however, Hucker offers unverified dates of 2205–1766 B.C.E.

60. Duan, Gan, Wang, and Chien, "Relocation of Civilization Centers," 572.

61. Ibid., 573.

62. Elvin, "Environmental Legacy of Imperial China," 743.

63. Ibid., 742–43.

64. Jinxiong Xu, *Zhongguo Gudai Shehui* (Society in Ancient China) (Taipei: Taiwan shangwu, 1988), 408–11, data cited in Elvin, "Environmental Legacy of Imperial China," 740.

65. Huang, *China: A Macro History,* 25.

66. Ibid., 24.

67. Ibid., 25–26.

68. Fang, "Chinese Perceptions," 46.

69. Qu and Li, *Population and Environment,* 17.

70. Archaeologists in Gansu Province in 2002 discovered an imperial order dating to the Han that banned felling trees and hunting young

animals in spring, burning wood in summer, and mining in autumn. "China Finds Oldest Environmental Protection Rule," *People's Daily Online,* April 24, 2002, fpeng.peopledaily.com.cn/200204/24/print20020424_94647.html.

71. Qu and Li, *Population and Environment,* 17.
72. Elvin, "Environmental Legacy of Imperial China," 736–37.
73. Fang, "Chinese Perceptions," 69.
74. Qu and Li, *Population and Environment,* 21.
75. Ibid., 22.
76. Ibid., 24–25.
77. Yuqing Wang, "Natural Conservation Regions in China," *Ambio* 16, no. 6 (1987): 326.
78. Elvin, "Environmental Legacy of Imperial China," 736.
79. Ibid., 747.
80. Jasper Becker, *Hungry Ghosts* (New York: Henry Holt, 1996), 11.
81. Angang Hu and Ping Zou, *China's Population Development* (Beijing: China's Science and Technology Press, 1991), 60.
82. Elvin, "Environmental Legacy of Imperial China," 753.
83. Dunstan, "Official Thinking on Environmental Issues," 592.
84. Ibid., 592.
85. Hu and Zou, *China's Population Development,* 60.
86. Qu and Li, *Population and Environment,* 16.
87. Ibid.
88. Ibid., 24–25.
89. Ibid., 23.
90. Dunstan, "Official Thinking on Environmental Issues," 592.
91. Ibid.
92. T'ung Tsu Chu, *Local Government in China under the Ch'ing* (Cambridge: Harvard University Press, 1988), 116–67.
93. Ch'ing hui-tien shih-li, chüan 927 in *Local Government in China under the Ch'ing* 167.
94. Dunstan, "Official Thinking on Environmental Issues," 591.
95. Ibid., 602.
96. Sen Yu, "On the Growing of Trees," in *Statecraft Anthology,* j.37, p. 11a–b, quoted in Dunstan, "Official Thinking on Environmental Issues," 602.
97. Dunstan, "Official Thinking on Environmental Issues," 603–5.
98. Ibid., 605–6.
99. Ibid., 595.
100. Ibid., 609.
101. Ibid., 609–10.
102. Mary B. Rankin, John K. Fairbank, and Albert Feuerwerker, "Intro-

duction: Perspectives on Modern China's History," in *The Cambridge History of China*, ed. Denis Twitchett and John K. Fairbank, vol. 13, *Republican China 1912–1949*, part 2, ed. John K. Fairbank and Albert Feurerwerker (Cambridge: Cambridge University Press, 1986), 6–7, 15.

103. Ibid., 7.
104. Ramon H. Myers, "The Agrarian System," in *Cambridge History of China*, 13:257.
105. Ibid., 13:256.
106. Lloyd E. Eastman, "Nationalist China during the Nanking Decade 1927–1937," in *Cambridge History of China*, 13:152.
107. Ibid., 13:151.
108. Lester Ross, *Environmental Policy in China* (Bloomington: Indiana University Press, 1988), 37.
109. In formulating his strategy to restore China to its previous greatness, Mao also drew inspiration from the experience of his imperial predecessors, especially Emperor Zhou of the Shang dynasty and Qin Shihuangdi, first emperor of the Qin dynasty. Both were notable for their success in expanding the territory of China and for their ruthless consolidation of power. (Zhisui Li, *The Private Life of Chairman Mao* [New York: Random House, 1994], 122.)
110. Mao Zedong, "Essay on How Shang Yang Established Confidence by the Moving of the Pole," in Stuart R. Schram, *Mao's Road to Power: Revolutionary Writings 1912–1949* (Armonk, N.Y.: M. E. Sharpe, 1992), 5.
111. Mao, "Essay on How Shang Yang Established Confidence," 6.
112. Judith Shapiro, *Mao's War against Nature* (Cambridge: Cambridge University Press, 2001), 8.
113. Mao Tse-tung [Mao Zedong], "Speech at the Inaugural Meeting of the National Science Research Society of the Border Region," February 5, 1940, in *Quotations from Mao Tse-tung* (Peking: Foreign Languages Press, 1966), 204–5.
114. Murphey, "Man and Nature," 319.
115. Ibid., 319.
116. Ibid., 320.
117. "Communique of the 11th Session of the Eighth Party Central Committee, August 14, 1966," in Hu and Zou, *China's Population Development*, 103.
118. *People's Daily* (January 1, 1973), in Hu and Zou, *China's Population Development*, 103.
119. Murphey, "Man and Nature," 323.
120. Hu and Zou, *China's Population Development*, 67.

121. Mao Tse-tung, "The Bankruptcy of the Idealist Conception of History," in *Selected Works of Mao Tse-tung* (Peking: Foreign Languages Press, 1967), 4:453.

122. Hu and Zou, *China's Population Development*, 104.

123. Ibid., 68 (1950 population); Geping Qu, *Environmental Management in China* (Beijing: China Environmental Science Press, 1991), 218 (1976 population); Fifth National Census, released March 28, 2001, www.macrochina.com.cn/english/china/resources/2001041 2000135.shtml (current population).

124. Shapiro, *Mao's War against Nature*, 33.

125. Hu and Zou, *China's Population Development*, 88.

126. Ibid.

127. Shapiro, *Mao's War Against Nature*, 45.

128. Hu and Zou, *China's Population Development*, 105.

129. Qu, *Environmental Management in China*, 211–12.

130. Mao Zedong, "Request for Opinions on the Seventeen-Article Document Concerning Agriculture," *Selected Works of Mao Tse-Tung* (Beijing: Foreign Languages Press, 1977): 5:279.

131. Becker, *Hungry Ghosts*, 70–82.

132. Vaclav Smil, *The Bad Earth: Environmental Degradation in China*, quoted in Becker, *Hungry Ghosts*, 77.

133. Becker, *Hungry Ghosts*, 77.

134. Mao Zedong, "Speech at the Fifteenth Meeting of the Supreme State Council, 8 September 1958 [excerpt]," Cold War International History Project, Woodrow Wilson International Center for Scholars, wwics.si.edu.

135. Becker, *Hungry Ghosts*, 63–64.

136. Antoaneta Bezlova, "Environment-China: Beijing Gasps for Clean Air," Inter Press Service, January 3, 2000.

137. Qu, *Environmental Management in China*, 212.

138. Shapiro, *Man's War against Nature*, 82–83.

139. Qu, *Environmental Management in China*, 212.

140. Shapiro, *Mao's War against Nature*, 89.

141. Hu and Zou, *China's Population Development*, 13.

142. Qu and Li, *Population and Environment*, 28.

143. Qu, *Environmental Management in China*, 213.

144. Ibid.

145. Murphey, "Man and Nature," 330.

146. Smil, *The Bad Earth*, 16.

147. Qu, *Environmental Management in China*, 213.

Chapter 3. The Economic Explosion and Its Environmental Cost

1. Susan Shirk, *The Political Logic of Economic Reform in China* (Berkeley: University of California Press, 1993), 176–81.
2. Ibid., 181–82.
3. Asian Development Bank, "Key Indicators 2002: Population and Human Resource Trends and Challenges," www.adb.org/Documents/ Books/Key_Indicators/2002/prc.pdf; "United Nations Statistics Division, National Accounts Main Aggregates Database, China, July 2004 Update," http://unstats.un.org/unsd/snaama/resultsCountry.asp? Country=156&SLevel=0&Year=2003&x=35&y=14&Selection=quick.
4. Kate Xiao Zhou, *How the Farmers Changed China* (Boulder, Colo.: Westview Press, 1996), 46–71.
5. "2000 Niandu Quanguo Xiangzhen Qiye Tongji Gong Bao," (2000 National Statistical Report on Township and Village Enterprises), January 1, 2001, Bureau of Township Enterprise Online, www.cte .gov.cn/zw/tjxx/readxx.asp?idd=31.
6. Dai Yan, "Foreign Investment Hits Record," *China Daily*, January 5, 2003, www.chinadaily.com.cn/cndy/2003-01-15/101359.html.
7. Ron Duncan and Xiaowen Tian, "China's Interprovincial Disparities: An Explanation," China Center for Economic Research, Working Paper Series, no. 1999012, table 2.
8. "China Basic Statistics by Province, 2002," www.ers.usda.gov/ Briefing/China/Data/tbl-provinces.xls; "China's Foreign Direct Investment Hits US$ 53 Bln in 2003," *People's Daily Online*, January 15, 2004, http://english.people.com.cn/200401/14/eng20040114_132611 .shtml.
9. Joshua Muldavin, "The Paradoxes of Environmental Policy and Resource Management in Reform-era China," *Economic Geography* 76, no. 3 (2000): 255.
10. *China Daily*, July 18, 1994.
11. *China Daily*, January 22, 2001.
12. Vaclav Smil, *China's Environmental Crisis: An Inquiry into the Limits of National Development* (Armonk, N.Y.: M. E. Sharpe: 1993), 61.
13. *World Resources 1994–1995* (New York: Oxford University Press, 1994), 131.
14. John Pomfret, "China's Lumbering Economy Ravages Border Forests; Logging Industry Taps Unregulated Markets for Wood," *Washington Post*, March 26, 2001, p. A19.
15. Brook Larmer and Alexandra A. Seno, "A Reckless Harvest," *The*

Bulletin, January 22, 2003, www.bulletin/EdDesk.nsf/All/017D7 A2BB84250155ECA256CB400071927.

16. The official estimate of total forestland in China is 16.55 percent (*2003 Nian Zhongguo Huanjing Zhuangkuang Gongbao* [2003 Report on the State of the Environment in China] [Beijing: State Environmental Protection Administration, 2004]) or approximately 159 million hectares. According to the 1998 and 2000 State of the Environment reports from SEPA, the amount of forested land expanded from 13.92 percent to 16.55 percent in those two years (SEPA, *State of the Environment Report*, 1998, 2000, www.zhb.gov.cn/english/ SOE/index.htm.) Since the beginning of China's history, China has lost approximately twice that amount—about 290 million hectares of forest or an area three times the size of Alaska. ("Tougher Measures Urged to Enforce Law," *China Daily*, August 30, 1999, p. 2).

17. SEPA, *State of the Environment Report*, 2001, www.zhb.cn/ english/soe/soechina2001/english/2-forest.htm.

18. Bochuan He, *China on the Edge* (San Francisco: China Books and Periodicals, 1991), 25–26.

19. Sichuan Environmental Protection Bureau official, taped interview with author's research associate, June 14, 2000, transcript on deposit with author.

20. *China Human Development Report 2002: Making Green Development a Choice* (Stockholm Environment Institute and UNDP China, 2002), 19.

21. "Tougher Measures Urged to Enforce Law," *China Daily*, August 30, 1999, p. 2.

22. Daniel Katz (founder, Rainforest Alliance), telephone conversation with author, March 5, 2002.

23. State of the Environment Reports, 1996, 1997, 1998, 1999, 2000, 2001, 2002.

24. World Bank, *China: Air, Land, and Water* (Washington, D.C.: World Bank, 2001), 23, lnweb18.worldbank.org/eap/eap.nsf/Attachments/ China+Env+Report/$File/China+Env+Report.pdf.

25. Qu and Li, *Population and Environment*, 61–62.

26. Ron Gluckman, "The Desert Storm," *AsiaWeek*, October 13, 2000, www.asiaweek.com/asiaweek/magazine/2000/1013/is.china.html.

27. Shi Yuanchun, "Reflections on Twenty Years' Desertification-control," www.china.com.cn/english/2002/May/32353.htm.

28. Qu and Li, *Population and Environment*, 76.

29. David Murphy, "Desertification—To Heal a Barren Land," *Far Eastern Economic Review*, July 19, 2001, p. 30.

30. Lester R. Brown, "Dust Bowl Threatening China's Future," *Earth Policy Alert*, May 23, 2001, p. 1.
31. John Copeland Nagle, "The Missing Chinese Environmental Law Statutory Interpretation Cases," *New York University Environmental Law Journal* (1996): 520.
32. Gareth Cook, "Massive Dust Cloud to Travel over the N.E.," *Boston Globe*, April 20, 2001, p. B4.
33. World Bank, *China: Air, Land, and Water*, 32.
34. Ibid., 32–35.
35. As defined by the World Bank in its *1992 World Development Report* (New York: Oxford University Press, 1992, p. 48), water scarcity is a severe constraint when annual renewable water resource supply is less than 1,000 m³ per capita. Water scarcity exists, however, when there is less than 2,000 m³ per capita.
36. "China's Water Shortage to Hit Danger Limit in 2030," Xinhua News Agency, November 16, 2001.
37. "World Water Day," www.worldwaterday.org/1997/cel.html.
38. *China Development Report 2002*, 24.
39. Jonathan Lassen, "Thirsty Cities and Factories Push Farmers Off Parched Earth," *China Development Brief* 4, no. 1 (2001): 13.
40. Population Action International, *Sustaining Water: Population and the Future of Renewable Water Supplies* (Washington, D.C.: Population Action International, 1993), www.cnie.org/pop/pai/h2o-toc .html.
41. By far the greatest proportion of water resources is consumed by agriculture. China consumes about 460 billion m³ of water annually, of which 87 percent is used for agriculture, 6 percent for household purposes, and 7 percent for industry. ("Water Everywhere: Is there Enough to Drink?" May 2002, www.actionbioscience.org/environment/lessons/kassasslessons.pdf.)
42. *China Statistical Yearbook*, 1988 and 1998 editions (Beijing: China Statistical Publishing House).
43. *China Statistical Yearbook*, 1988–1994 editions (Beijing: China Statistical Publishing House).
44. *China News Analysis*, December 1, 1993, p. 5.
45. Philip P. Pan, "Wetlands Running Dry in China; Drought Erodes an Ancient Way of Life in Mythic Marshes," *Washington Post*, July 1, 2001, p. A14.
46. SEPA, Report on the State of the Environment in China, 1999, www.zhb.gov.cn/english/SOE/soeChina1999/index.htm.
47. "Water Sources for China Yellow River Dry Up," Xinhua News

Agency, October 9, 2001, www.planetarn.org/dailynewsstory.ctm/newsid/12711/story.htm.

48. "China's Water Shortage to Hit Danger Limit in 2030," Xinhua News Agency, November 16, 2001.

49. Lassen, "Thirsty Cities," 13–14.

50. "Turn Off Tap to Firms That Waste Water," *China Daily*, March 4, 2003, wwwl.chinadaily.com.cn/chinagate/doc/2003-03/04/content_247320.htm.

51. *2003 Nian Zhongguo Huanjing Zhuangkuang Gongbao (*2003 Report on the State of the Environment in China).

52. Changhua Wu, Crescencia Maurer, Yi Wang, Shouzheng Xue, and Dewa Lee Davis, "Water Pollution and Human Health in China," *Sinosphere* 2, no. 3 (1999): 12.

53. SEPA, *2002 State of the Environment Report*, www.zhb.gov.cn/English/SOE/soechina2002/index.htm; *2003 Nian Zhongguo Huanjing Zhuangkuang Gongbao* (2003 Report on the State of the Environment in China).

54. The main sources of China's water pollution are industrial and municipal wastewater discharges, agricultural runoff from chemical fertilizers, pesticides and animal manure, and the leaching of solid waste. (World Bank, *Clear Water, Blue Skies*, 90.)

55. *China Human Development Report*, 33.

56. Vaclav Smil, "China Shoulders the Cost of Environmental Change," *Environment* 39 (July/August 1997): 9, 33.

57. *China Environment News*, October 4, 1994, p. 4.

58. In one early 1990s case, for example, in a rural village in Hebei Province, farmers established a tannery that earned them revenues of 300 million yuan ($36.6 million). However, in 1993, this tannery discharged 11.3 million m^3 of wastewater with a high content of sulfides and chromium directly into sewage pits. This wastewater seriously damaged surface and groundwater, reduced crop yields, and produced "sour" fruit. However, the farmers claimed they were indifferent to the poor yield because the tannery was far more important to their economic well-being. (*China Environment News*, October 4, 1994, p. 4.)

59. *China Human Development Report*, 25.

60. *China Statistical Yearbook 2002* (Beijing: China Statistical Publishing House, 2002), 389.

61. World Bank, *China: Air, Land, and Water*, 58.

62. "Beijing Air Quality Improves in 1999," Zhongguo Xinwen She News Agency [in Chinese] (Beijing: December 28, 1999), translated in *BBC Worldwide Monitoring* (February 2, 2000).

63. Frank Langfitt, "China Struggles to Clear Its Air," *Baltimore Sun*, January 17, 2000, p. 2A.
64. World Bank Group, *2001 World Development Indicators* (Washington, D.C.: World Bank, 2001), 174; *China Development Report*, 26.
65. "Pollution Control Reinforced," *China Daily*, May 4, 1999, p. 2.
66. "Acid Rain Causes Annual Economic Loss of 110 Billion Yuan in China," *China Daily*, October 10, 2003.
67. Energy Information Administration, 1999 figures.
68. *China Statistical Yearbook 2002*, 249.
69. Ibid., 28–29.
70. Chenggang Wang, "China's Environment in the Balance," *World and I* 14 (October 1999): 176.
71. Overall, the percentage of coal in the energy mix for China's commercial energy needs has remained relatively constant at 75 percent. Ibid.
72. Coal combustion in industrial boilers and small household stoves (most households still burn solid fuels such as raw coal and wood for cooking and heating) account for up to two-thirds of ambient levels of fine and ultrafine particles, which are the most damaging to human health. These boilers—which are usually inefficient and emit through low smokestacks (*World Resources 1998–1999* [New York: Oxford University Press, 1998], 117)—and stoves are also responsible for most sulfur dioxide and nitrogen oxide emissions. Over time, China has been replacing these boilers with more efficient ones, however, the less efficient boilers are often maintained in service by industry outside the major cities.
73. *China Human Development Report*, 32.
74. Much of this coal has high sulfur content and impurities, sometimes as much as 30 to 45 percent of its total weight. This has contributed to both serious local air quality problems as well as acid rain.
75. Daniel Esty and Robert Mendelsohn, "Powering China: The Environmental Implications of China's Economic Development" (unpublished manuscript, Yale Center for Environmental Law and Policy, 1995), 8.
76. The government has pushed to close down many of these small coal mines; according to one report in Gansu Province, a major coal consumer, officials closed down 648 such mines in 1999, roughly half of the 1,300 run by township enterprises or state firms. The motive for closing the mines appears to have been a combination of (1) ending the oversupply of coal and price war on coal that had emerged and (2) cutting down on pollution and poor safety ("Northwestern China Province Closes 648 Small Coal Mines," Xinhua News Agency, Janu-

ary 27, 2000). This supports Asian energy expert Robert Manning's finding that by 1999, the percentage of coal in the overall energy mix had dropped precipitously to 68 percent from 75 percent as a result of "overcapacity, lower growth, and shutting down of a host of mines" as well as the "trend toward increased gas use." Manning also argues, however, that the drop in coal use is a short-term phenomenon. (Robert Manning, *The Asian Energy Factor*, New York: Palgrave, 2000.)

77. Hilary French, "Can Globalization Survive the Export of HAZARD," *USA Today*, May 1, 2001, p. 23.

78. Yulanda Chung "Now You See It . . . Now You Don't," *Asiaweek*, October 1, 1999, www.asiaweek.com/asiaweek/magazine/99/1001/hongkong.html.

79. Ibid.

80. *Exporting Harm: The High-Tech Trashing of Asia*, ed. Jim Puckett and Ted Smith (The Basel Action Network and Silicon Valley Toxics Coalition, February 25, 2002), 22, www.svtc.org/cleancc/pubs/technotrash.pdf.

81. Cheung Chi-fai, "Greens Fear Impact of Cross-Border Bridge Project," *South China Morning Post*, October 2, 2001, p. 4.

82. Phil Murtaugh, "Why China? Trends and Opportunities in the PRC," May 22, 2002, www.autonews.com/files/Murtaugh.ppt.

83. *China Human Development Report*, 28.

84. *China Statistical Yearbook 2002*, 550–52.

85. Jasper Becker, "Car-Crazy Mainland Risks a Wrong Turn," *South China Morning Post*, December 28, 2001, p. 12.

86. During the mid-1990s, the government set out new exhaust emission standards to reduce carbon dioxide and hydrocarbon emissions by thirty times, also mandating newly designed carburetors and the use of catalytic converters to help facilitate reduction of hydrocarbons and nitrogen oxide. However, even as Chinese auto makers have undertaken technical renovations to improve the level of auto emissions, they have complained to the government that no matter what technical improvements are made they will not be able to meet the national standards because of the low quality of domestically refined fuels. They argue that the "low-quality fuel will cause an engine's efficiency to deteriorate quickly and an exhaust filter to lose its effectiveness." The car manufacturers have appealed to the government to import foreign fuel in order to meet the national standards. ("China—Automakers Urge China to Import Unleaded Gasoline," *China Online*, August 31, 2000).

87. Becker, "Car-Crazy Mainland," p. 12.

88. *World Resources 1998–1999,* 118.

89. Hu and Zou, *China's Population Development,* 92.

90. Ibid., 92–93

91. "Jiang Says China's Population His Biggest Problem," www. Dateline.muzi.net/11/english/46389.shtml.

92. Zhou, *How the Farmers Changed China,* 181.

93. Ibid., 190–93.

94. People's Health Publishing House, "China's Family Planning Yearbook," in Hu and Zou, *China's Population Development,* 10.

95. Jiang Zhuzing, "Population Growth Well Under Control," *China Daily,* March 29, 2001.

96. Zhou, *How the Farmers Changed China,* 176.

97. Chinese economist He Qinglian and her Princeton University associate Cheng Xiaonong argue that the reported average family size of 3.44 must be wrong: "City people comprise only 30% of the population, people in the countryside can have two children and many do." Qinglian He and Xiaonong Cheng, "Population, the Economy, and Resources" (summary of article at: www.usembassy-china.org.cn/Sand+/heqinglian-populationresources.html).

98. Matthew Miller, "Writing's on the Wall for Rural China," *South China Morning Post,* May 10, 2001, p. 17.

99. He and Cheng, "Population, the Economy, and Resources."

100. United Nations Economic and Social Commission for Asia and Pacific, "Sustainable Social Development in a Period of Rapid Globalization: Challenges, Opportunities, and Policy Options" (UNESCAP, 2002).

101. He and Cheng, "Population, the Economy, and Resources."

102. In 1994, China bought 34 percent more grain—including wheat, corn, and rice—than in 1993, an increase of 16 million tons. In 1994–1995, Chinese wheat buying increased by 13 percent to 17.9 million tons; and in 1995–1996, imports dropped only slightly to 16.25 million tons. (Lester Brown, *Who Will Feed China?* [New York: W.W. Norton, 1995], 100.) By 1995, China had become the world's second largest grain importer, second only to Japan.

103. Xie was echoing the title of Lester Brown's article, which was expanded into a full-length book. Brown, *Who Will Feed China?* 17.

104. Guangdong, in particular, was criticized for neglecting its agricultural base; it routinely suffered a shortfall of 2 million tons. ("Commentary Views Agricultural Problem," Xinhua News Agency, February 25, 1995, in Foreign Broadcast Information Service [hereafter FBIS], China Daily Report, March 6, 1995, p. 81.)

105. "Party Chief Discusses Importance of Agriculture," Xinhua News Agency, February 27, 1995.
106. "Renmin Ribao on Conference," *People's Daily,* February 28, 1995, in FBIS, China Daily Report, March 1, 1995, p. 64.
107. *China Statistical Yearbook 1999,* 378.
108. "Severe Drought in Northern China Threatens Agricultural Output for 2nd Year," *Dow Jones International News,* May 30, 2001.
109. "Drought-hit Chinese Farmers Protest over Fees Policy," *Kyodo News,* June 13, 2001.
110. Zhang Wenjie, "Grain Safety," *cctv.com,* February 4, 2004, http://www.cctv.com/english/special/chinatoday/20040204/100699.shtml.
111. "Provincial Agriculture Officials on Investment," *People's Daily,* February 26, 1995, in FBIS, China Daily Report, February 26, 1995, p. 60.
112. Joseph Kahn, "Feeding the Masses: China's Industrial Surge Squeezes Grain Farms, Spurs Needs for Imports," *Wall Street Journal* (Europe), March 13, 1995, p. A6.
113. Lester Brown, "Dust Bowl Threatening China's Future," 2.
114. Ibid.
115. "China Agriculture: Cultivated Land Area, Grain Projections, and Implications" (MEDEA summary report, November 1997).
116. According to the SEPA 2000 report, there has been a threefold increase in the use of agricultural chemicals since 1995.
117. *China Human Development Report,* 20.
118. "China Reports Erosion Reducing Black Soil in Heilongjiang Province," Xinhua News Agency, January 25, 2000, collected by *BBC Worldwide Monitoring.*
119. Zhong Ma, "China's Past Speaks to Sustainable Future," *Forum for Applied Research and Public Policy* 10 (winter 1995): 53–54.
120. These include both those who have registered as temporary residents and those who have not. The size of the latter has been expanding since the early 1980s. (Laurence J. C. Ma and Biao Xiang, "Native Place, Migration, and the Emergence of Peasant Enclaves in Beijing," *China Quarterly* 155 [September 1998]: 556–587.)
121. Ibid., 554–55.
122. Vaclav Smil, "China's Environmental Refugees: Causes, Dimensions, and Risks of an Emerging Problem," in *Environmental Crisis: Regional Conflicts and Ways of Cooperation, Report of the International Conference at Monte Verita Ascona, Switzerland, 3–7 October 1994* (Center for Security Studies, 1995), 83.
123. Smil, "China's Environmental Refugees," 86.
124. The Chinese leadership is reviewing proposals, already in an exper-

imental stage, to change the system of residence requirements and permit the migrant population and the rest of China's population to change their residence status. Still, limits will remain on cities such as Shanghai and Beijing that already feel overburdened by their migrant populations. Instead, the minister of civil affairs announced a plan in 2000 to build four hundred new cities, at the rate of around twenty per year, up to 2020 to increase the mainland's urban population to 800 million. Shanghai, alone, is planning for ten new cities outside Shanghai, with an eye toward attracting "above average income" families who will move to the cities and create new businesses and work opportunities for farmers. (Geoffrey Murray, "Creating New Cities Helps China Cope with Migrants," *Kyodo News*, March 23, 2001.)

125. Ching-ching Ni, "17 Women Are Key to Link in Shanghai's Outdated Sewage System," *Los Angeles Times*, March 4, 2001, p. A3.

126. Damien McElroy, "China to Resettle 300 Million: Leaders Hope Five-Year Migration into New Towns Will Boost Economy," *Ottawa Citizen*, November 14, 1999.

127. Lorien Holland, "Running Dry," *Far Eastern Economic Review*, February 3, 2000, p. 18.

128. Ibid., 19.

129. "Drought-hit Chinese Farmers Protest over Fees Policy," *Kyodo News*, June 13, 2001.

130. Independent of such a tension, the challenge would be unlikely to arise from the migrants themselves. They tend to be politically powerless, clustering in villages within the cities that replicate their own provincial background, and without the opportunity thus far to establish real roots in the society or become legitimate citizens. (Ma and Xiang, "Native Place, Migration, and the Emergence of Peasant Enclaves," 578–79.)

131. Dexter Roberts "The Great Migration: Chinese Peasants Are Fleeing Their Villages to Chase Big-City Dreams," *Business Week* (international ed.), December 11, 2000, www.businessweek.com/datedtoc/2000/0050t.htm.

132. Ibid.

133. Yan Cui, Zuo-Feng Zhang, John Froines, Jinkou Zhao, Hua Wang, Shun-Zhang Yu, and Roger Detels, "Air Pollution and Case Fatality of SARS in the People's Republic of China: An Ecologic Study," *Environmental Health: A Global Access Science Source* 2, no. 15 (November 20, 2003): n.p.; "China tightens control over disposal of SARS-related medical waste," www.english.enorth.com.cn/system/2003/04/29/000552829.shtml.

134. *China Human Development Report,* 33.

135. James Kynge, "Yellow River Brings Further Sorrow to Chinese People," *Financial Times,* January 7, 2000, p. 8.

136. Philip Pan, "Wetlands Running Dry in China," *Washington Post,* July 1, 2001, p. A14.

137. Lassen, "Thirsty Cities," 14.

138. "Focus: Global Environmental Situation Depends on China," *Kyodo News,* February 28, 2002.

139. Huang et al., "Farm Pesticide, Rice Production, and Human Health," Center for Chinese Agricultural Policy, Chinese Academy of Agricultural Sciences Project Report 11 (2000).

140. Particulates and sulfur dioxide contribute to chronic obstructive pulmonary disease, lung cancer, heart disease, and stroke; increased risk of respiratory infection; and impaired lung functioning. A study of the health effects of air pollution in Shenyang and Shanghai indicated that indoor air pollution, primarily from coal burning stoves, was responsible for 15 to 18 percent of lung cancers and outdoor pollution contributed to another 8 percent.

141. Hua Wang, Jun Bi, and David Wheeler et al., "Environmental Performance Rating and Disclosure: China's Green-Watch Program," World Bank Policy Research Working Paper 2889, September 2002, p. 4 (www.econ.worldbank.org/files/18746_wps2889.pdf.)

142. Ivan Tang, "Mainland Air Deadly for Children," *South China Morning Post,* May 6, 1999.

143. While China has begun to phase out leaded gasoline, in Beijing "30% of the gas used is estimated to fall short of the government's environmental standards." Henry Chu, "China Is Passing Pollution to a New Generation, Study Finds," *Los Angeles Times,* June 19, 2002.

144. Ibid.

145. Caixiong Zheng, "Children Have Too Much Lead in Their Bloodstreams," *China Daily,* August 18, 2000.

146. *China Human Development Report,* 28.

147. George Wehrfritz, "Green Heat," *Newsweek,* October 7, 1996, 16.

148. Ibid.

149. Yong Huang, "You Du Dami Shei Gan Ru Kou" (Who Dares Eat the Poisoned Grain), *Zhongguo Huangjing Bao* [China Environment News], November 1, 2001, www.ee65.com.cn:8080/cgi-bin/CgiSrch. exe?FID=ZjZZCdNI&OID=5.

150. Jing Jun, "Environmental Protests in Rural China," in *Chinese Society: Change Conflict and Resistance,* ed. Elizabeth J. Perry and Mark Selden (New York: Routledge, 2000), 144.

151. Melinda Liu, "Waiting for Rain: China Is Struggling to Cope with Its Worst Drought in 50 Years," *Newsweek*, August 13, 2000.

152. Jing Jun, "Environmental Protests," 146–47.

153. Jing Jun, interview with author, New York, December 5, 2001.

154. Ibid.

155. Erik Eckholm, "China's Inner Circle Reveals Big Unrest, and Lists Causes," *New York Times*, June 3, 2001, p. 14.

156. *Inside China Mainland* 19, no. 3 (1997): 94, quoted in Lisa Eileen Husmann, "Falling Lands, Rising Nations: Environmental-Nationalism in China and Central Asia" (Ph.D. diss., University of California, Berkeley, 1997), 150.

157. Jing Jun, "Environmental Protests," 157.

158. The real numbers are likely much higher once the damage from large-scale disasters such as the 1998 Yangtze River floods or the 2001 drought are included.

159. World Bank, *Clear Water, Blue Skies*, 23.

160. Ibid.

161. "Pollution causing significant losses to China's fishing industry," *AFX News*, June 24, 2001.

162. Environment, Science, and Technology Section, U.S. Embassy Beijing, "Issues Surrounding China's South-North Water Transfer Project" (April 2001), www.usembassy-china.org.cn/english/sandt/SOUTH-NORTH.html.

163. "Drought Leaves China Dry and High," *CNN.com* (June 20, 2001), www.cnn.com/2001/WORLD/asiapcf/east/06/20/china.tapsoff.

164. Tamar Hahn, "China Dealing with a Wealth of Environmental Challenges," *The Earth Times*, March 27, 2001.

Chapter 4. The Challenge of Greening China

1. Geping Qu, *Environmental Management in China* (Beijing: China Environmental Sciences Press, 1999), 214.

2. Ibid.

3. United Nations, "Report of the United Nations Conference on the Human Environment" (paper presented in Stockholm, 1972) [hereafter UNCHE], www.unep.org/Documents/Default.asp?DocumentID=97.

4. "Chinese Delegation Makes Statement on 'Declaration on Human Environment,'" *Peking Review*, June 23, 1972, pp. 9–11.

5. UNCHE, "Report of the United Nations Conference on the Human Environment," UN Document A, conf.48/14/Rev.1.

6. This leading group included officials from the ministries of planning, industry, agriculture, communications, water conservancy, and public health.
7. Qu, *Environmental Management in China,* 214–16.
8. Ibid., 219.
9. Ibid., 219.
10. Ibid., 222.
11. Abigail Jahiel, "The Organization of Environmental Protection in China," *China Quarterly* (December 1998): 769.
12. Ibid., 150.
13. Xiaoying Ma and Leonard Ortolano, *Environmental Regulation in China* (Lanham, Md.: Rowman & Littlefield, 2000), 16.
14. Qu, *Environmental Management in China,* 113–14.
15. State Science and Technology Commission official, interview with author, Beijing, April 1992.
16. During the UNCED, countries engaged in the final negotiations for a framework convention on climate change to control the emissions of greenhouse gases. One of the top emitters of these gases, along with the United States and Russia, China categorically refused to consider any targets or timetables for limiting its emissions, arguing that, as a newly industrializing country, it bore little historical responsibility for the problem of climate change and that given its status as a developing country, it should not be expected to take action unless fully compensated by the advanced industrialized nations.
17. A commonly accepted definition of *sustainable development* is provided by the World Commission on Environment and Development in *Our Common Future* (Oxford: Oxford University Press, 1987), 43: "The ability of humanity to ensure that it meets the needs of the present without compromising the ability of future generations to meet their own needs. Sustainable development is not a fixed state of harmony but rather a process of change in which the exploitation of resources, the direction of investments, the orientation of technological development and institutional changes are made consistent with future as well as present needs."
18. In the aftermath of the UNCED, for example, China became the first country to develop an action plan embodying the ideal of sustainable development. China's Agenda 21, modeled on the UNCED's global Agenda 21, was a call to action on every environmental issue from local air pollution to biodiversity. The program incorporated input from over 300 Chinese ministries, commission and local govern-

ments, and totaled 128 projects. (SEPA official, interview with author, Beijing, April 2000.) Within the Priority Program for China's Agenda 21, for example, there was also a call to enhance the legal system in order to attain sustainable development. This led to revisions in laws such as the Air Pollution Prevention and Control Law in 1995 (ibid.).

19. This was stated in the "Notice on strengthening management of environmental impact assessments in construction projects undertaken with loans from international financial organizations." Qing Dai and Eduard B. Vermeer, "Do Good Work, But Do Not Offend the 'Old Communists,'" in *China's Economic Security*, ed. Robert Ash and Werner Draugh (New York: St. Martins, 1999), 143.

20. Ibid.

21. In the wake of the UNCED, international environmental NGOs began to flood into China with much-needed technical expertise and funds. International business, too, became more directly involved in elevating the environmental awareness of their Chinese counterparts and in contributing to development projects under the rubric of sustainable development.

22. *World Resources 1998–1999* (Oxford: Oxford University Press, 1998), 124.

23. Annette Chiu, "Future Looks Grim, Green Chief Admits," *South China Morning Post*, April 9, 2001.

24. According to Alford and Liebman, "The NPC's power include the authority to enact all 'basic laws' (*jiben fa*), to supervise the implementation of such laws, and to make amendments to the Constitution. The full NPC meets only once a year . . . most law-making activity is instead conducted by its [roughly 155-member] Standing Committee, [which] is authorized to interpret the Constitution, pass laws (*fa*) other than basic laws, which are the domain of the full NPC, interpret laws, and supervise the work of the other principal organs of government." (William P. Alford and Benjamin L. Liebman, "Clean Air, Clear Processes? The Struggle over Air Pollution Law in the People's Republic of China," *Hastings Law Journal* 52 [March 2001]: 706–7.)

25. The name of the National Environmental Protection Agency was changed to the State Environmental Protection Administration in 1998, marking at the same time an elevation in the agency's status to the level of a ministry.

26. See, for example, Frank Ching, "Rough Justice: The Law Is No Longer an Ass, but Many Judges Still Are," *Far Eastern Economic Review*, August 20, 1998, p. 13; Jerome A. Cohen and John Lange, "The Chi-

nese Legal System: A Primer for Investors," *New York School of International and Comparative Law* 17 (1997): 345–77; Sam Hanson, "The Chinese Century: An American Judge's Observation of the Chinese Legal System," *William Mitchell Law Review* 28 (2001): 243–52.

27. Hongjun Zhang and Richard Ferris, "Shaping an Environmental Protection Regime for the New Century: China's Environmental Legal Framework," *Sinosphere* 1, no. 1 (1998): 6.

28. For a discussion of all the responsibilities of the EPNRC, see Zhang and Ferris, "Shaping an Environmental Protection Regime," 6.

29. Chun Jin, "New Laws to Ensure Better Environment," *China Daily*, November 20, 2000.

30. John Copeland Nagle, "The Missing Chinese Environmental Law Statutory Interpretation Cases," *New York University Environmental Law Journal* 5 (1996): 548–49.

31. Ma and Ortolano, *Environmental Regulation in China*, 92–93.

32. Ibid., 92.

33. "Environment Tax Beneficial," *China Daily*, January 23, 1999, p. 4.

34. EPNRC official, interview with author, Cambridge, Mass., January 1999.

35. Ibid.

36. Alford and Liebman, "Clean Air, Clear Processes?" 711.

37. Ibid., 714.

38. Without such a change, pollutants might be diluted or released slowly over time, with a significant impact on the environment but evading detection.

39. These included clean production technology, the sulfur content of coal, the establishment of acid rain control areas, and the use of unleaded gasoline.

40. Alford and Liebman, "Clean Air, Clear Processes?" 744–45.

41. Yan Meng, "Amended Law to Tackle Major City Air Pollution," *China Daily*, September 2, 2000.

42. "Key Chinese Cities Improve Air Quality," July 21, 2003, www.cenews.com.cn/english/2003-07-21/400.php.

43. "Environment Tax Beneficial," *China Daily*, June 23, 1999, p. 4.

44. Yong Zhang, "Tough Campaign Cleans Air," *China Daily*, September 26, 2001.

45. In contrast, the U.S. EPA headquarters boasts 6,000 employees. (Ferris and Hongjun, "Shaping an Environmental Protection Regime.")

46. Qu, *Environmental Management in China*, 221.

47. Hua Guo, "Nation's Plan to Protect Environment Outlined," *China Daily*, May 29, 2001.

48. *China Environment News,* July 1994, p. 1.

49. EPNRC official, interview with author, Cambridge, Mass., January 1999.

50. In 1998, the State Planning Commission was renamed the State Development and Planning Commission, and the State Science and Technology Commission was reorganized and renamed the Ministry of Science and Technology.

51. Agenda 21 was the global sustainable development action plan enunciated at the UNCED. Countries then developed their own Agenda 21 projects based on their individual needs and priorities.

52. World Bank Official, phone interview with author, November 2001. This issue is discussed in greater depth in chapter 7.

53. Jahiel, "Organization of Environmental Protection," 772.

54. *China Statistical Yearbook 2002* (Beijing: China Statistical Publishing House, 2002), 817.

55. Jahiel, "Organization of Environmental Protection," 759.

56. *China Environment News,* July 1994, p. 1.

57. Limin Liu, "Hebei Chachu Liangqi Huanjing Weifa Anjian Zeren Ren," [Hebei Takes Action against the Responsible Individuals in Two Cases], *Zhongguo Huanjing Bao* [China Environment News], December 21, 2001, www.ee65.com.cn:8080/cgi-bin/CgiSrch.exe? FID=pjYVCgtx&OID=5.

58. Ma and Ortolano, *Environmental Regulation in China,* 81.

59. Between 1985 and 1994, the number of firms that had charges levied against them grew from approximately 80,000 to over 300,000, and the total amount of fines levied reached approximately 3 billion yuan ($365 million) for pollution discharges. *China Environment News,* February 1995, p. 1.

60. This designation of *model environmental city* reflects a determination by SEPA that a city is effectively tackling its environmental problems based on five environmental quality indicators, ten indicators for urban environmental infrastructure, four indicators for environmental management, and four socioeconomic indicators.

61. Shanghai EPB official, interview with author, Shanghai, September 1999.

62. Baoping Wu, "Environmental Police in Shanxi," *East Environment,* January 14, 2002, www.ee65.com.cn/pub/gb/2001english/2002 .0114/20021.0114e.htm.

63. Guangzhou EPB official, interview with author, Guangzhou, April 2000.

64. Environment, Science, and Technology Section, U.S. Embassy Bei-

jing, "Eco-Recitivism [*sic*] Revisited," *Beijing Environment, Science and Technology Update,* August 10, 2001, www.usembassy-china. org.cn/english/sandt/estnews0810.htm.

65. A fee is levied only for pollutants that do not comply with regulatory effluent standards. No matter the number of pollutants a firm discharges that do not meet regulatory standards, the firm pays only the charge on the pollutant that exceeds the standard by the greatest amount. The fee collected supports the administration of the EPB, the discharge fee system, and the firm's pollution control projects. Up to 80 percent of the levy paid can be used to subsidize the pollution control project proposed by the firm. (Hua Wang, Nlandu Mamingi, Benoit Laplante, and Susmita Dasgupta, "Incomplete Enforcement of Pollution Regulation: Bargaining Power of Chinese Factories," working paper, World Bank, Washington, D.C., April 2002, p. 6.)

66. Abigail Jahiel, "Policy Implementation Under Socialist Reform: The Case of Water Pollution Management in the People's Republic of China" (Ph.D. diss., University of Michigan, 1994), 199.

67. Wang, Mamingi, Laplante, and Dasgupta, "Incomplete Enforcement of Pollution Regulation," p. 6.

68. Qu, *Environmental Management in China,* 318–19.

69. Wang, Mamingi, Laplante and Dasgupta, "Incomplete Enforcement of Pollution Regulation," p. 6.

70. Shanghai EPB official, interview with author, Shanghai, September 1999.

71. Jahiel, "Policy Implementation Under Socialist Reform," 97.

72. Ma and Ortolano, *Environmental Regulation in China,* 126.

73. Jasper Becker, "Cracking Down on the Chemical Waste Outlaws," *South China Morning Post,* May 21, 1999, p. 2.

74. Zou Shengwen, "Wo Guo Huanjing Jigou Jiang Bu Zai Chi Pai Wu Fei" [China's Environment Protection Apparatus Will No Longer Be Able to Pocket Waste Discharge Fees], January 15, 2002, www.china.org.cn/chinese/2002/Jan/97539.htm.

75. "Administrative Regulations on the Collection and Use of Pollutant Discharge Fees," in *China Law & Policy,* January 24, 2003, www.omm.com/webdata/content/publications/clp030124.pdf.

76. Nagle, "Missing Chinese Environmental Law," 523–24.

77. Jerome Cohen and John E. Lange, "The Chinese Legal System: A Primer for Investors," *New York Law School Journal of International and Comparative Law* 17 (1997): 350.

78. William P. Alford, "Limits of the Law in Addressing China's Envi-

ronmental Dilemma," *Stanford Environmental Law Journal* 16 (January 1997): 141.

79. Phyllis L. Chang, "Deciding Disputes—Factors That Guide Chinese Courts in the Adjudication of Rival Responsibility Conduct Disputes," *Law and Contemporary Problems* 52, no. 3, in Ma and Ortolano, *Environmental Regulation in China*, 91.

80. Nagle, "Missing Chinese Environmental Law," 537.

81. Ted Plafker, "Chinese Activists Take to the Courts," *International Herald Tribune*, August 28, 2002, p. 20.

82. Violence sometimes accompanies court rulings. In one 1994 case, after a county People's court in Hubei ordered a chemical plant to pay a fine it had avoided for two years, the workers became enraged, harassing local EPB officials and destroying their offices. (*China Environment News*, November 1994, p. 1.)

83. Nagle, "Missing Chinese Environmental Law," 530.

84. Chinese environmental lawyer, Council on Foreign Relations, interview with author, New York, January 1999.

85. Julie Chao, "More and More Chinese Saying 'I'll Sue' to Settle Disputes," *Cox Newspapers*, October 8, 2000, www.coxnews.com/ washingtonbureau/staff/chao/10–08–00CHINALAWSUISADV081S COX.html.

86. Frank Ching, "China: Rough Justice," 13.

87. C. David Lee, "Legal Reform in China: A Role for Nongovernmental Organizations," *Yale Journal of International Law* 25 (summer 2000): 367–68.

88. Shanghai EPB official, interview with author, Shanghai, September 1999.

89. Mark O'Neill, "Decorator Pays for Using Toxic Chemicals," *South China Morning Post*, January 2, 2003.

90. "Zhejiang Peasants Compensated for Pollution," www.cenews .com.cn/english/2003–01–13.

91. *China Environment News*, September 22, 2001, in "Beijing Environment, Science and Technology Update," October 5, 2001, www.usembassy-china.org.cn/sandt/estnews100501.htm.

92. *China Environment News*, February 19, 1995, p. 2.

93. *China Environment News*, May 15, 1995, p. 1.

94. Shanghai EPB official, interview with author, Shanghai, September 1999.

95. "China Hands Down First Sentence for Pollution," Agence France-Presse, October 9, 1998.

96. "Local Officials Sanctioned for Lax Environmental Enforcement," *China Daily*, August 31, 2001.

97. "Two Officials Imprisoned for Neglect of Their Duties," *China Daily*, August 6, 2002.

98. Wang Canfa, Council on Foreign Relations, interview with author, New York, May 18, 2001.

99. *China Environment News*, October 24, 2001.

100. Iain Johnson, "Mass Leverage: Class Action Suits Let the Aggrieved in China Appeal for Rule of Law," *Wall Street Journal*, March 25, 1999, p. A1.

101. Chao, "More and More Chinese Saying 'I'll Sue,' " 3.

102. The China Construction Bank, one of the four largest state-owned commercial banks, for example, reportedly raised $5.11 million in overseas funds for air pollution monitoring projects in eleven cities. These projects were the first concrete programs developed by China and the United States following President Clinton's 1998 visit to China. Eighty-five percent of the $5.11 million in funds came from a preferential loan from the Bank of America. ("Bank Clears Air with Help from US Loans," *China Daily*, August 19, 1999, p. 5.)

103. *China Environment News*, January 1995, p. 1.

104. The one case in which the bank played this role was mentioned by an EPB official in Dalian.

105. *China Environment News*, December 18, 2001.

106. The officials must sign contracts with local EPBs outlining specific environmental goals and pledging to work together to achieve these targets. Environmental protection is also, on paper, a criterion for future political advancement for these officials.

107. EPB officials, interviews with author, Shanghai (September 1999), Xiamen (April 2000), and Sichuan (June 2000).

108. Scott Rozelle, Xiaoying Ma, and Leonard Ortolano, "Industrial Wastewater Control in Chinese Cities: Determinants of Success in Environmental Policy," *Natural Resource Modeling* 7 (fall 1993): 362–68.

109. EPB officials, interviews with author's research associate, Chongqing and Kunming, June 2000.

110. EPB officials, interview with author's research associate, Chongqing, June 2000.

111. Susmita Dasgupta and David Wheeler, "Citizen Complaints as Environmental Indicators: Evidence from China," unpublished manuscript, Environment, Infrastructure and Agriculture Division, Policy Research Department, World Bank, Washington, D.C., 1996, in Ma and Ortolano, *Environmental Regulation in China*, 71.

112. EPB official, interview with author, Dalian, April 2000.
113. "Hotline to Help Settle Environmental Problems," July 30, 2001, www.xinhuanet.com/english/20010730/434912.htm.
114. EPB officials, interviews with author's research associate, Chongqing, Kunming, and Sichuan, June 2000.
115. "Chongqing Hotline Helps Monitor Pollution," China Westnews, November 21, 2002, www.cjmedia.com.cn/en/article/2002.
116. EPB officials, interview with author, Guangzhou, April 2000.
117. While not stated explicitly, the Shanghai EPB likely pushed for such a policy in response to efforts by NEPA and EPNRC to advance this policy at the national level, as detailed earlier. (EPB official, interview with author, Shanghai, September 1999.)
118. *World Resources 1998–1999*, 124.
119. "Drought Leaves China Dry and High," CNN.com, June 20, 2001, edition.cnn.com/2001/world/asiapcf/east/06/20/china.tapsoff.
120. Lorien Holland, "Running Dry," *Far Eastern Economic Review,* February 3, 2000, p. 19.
121. *Beijing Youth Daily,* January 19, 2003, in Environment, Science, and Technology Update, www.usembassy-china.org.cn/sandt/estnews013103.htm.
122. EPB official, interview with author, Dalian, April 2000.
123. EPB officials, interview with author, Guangzhou, April 2000.
124. EPB officials, interview with author, Zhongshan, April 2000.
125. *1997 China Statistical Yearbook* (Beijing: China Statistical Publishing House, 1997), 5; *China Statistical Yearbook 2002* (Beijing: China Statistics Press, 2002), 5.
126. John Pomfret, "China's Lumbering Economy Ravages Border Forests; Logging Industry Taps Unregulated Markets for Wood," *Washington Post,* March 26, 2001, p. A19.
127. Ibid.
128. Brook Larmer and Alexandra A. Seno, "A Reckless Harvest," *The Bulletin,* January 22, 2003, www.bulletin/eddesk.nsf/all/017D7A2BB842501ECA256CB400071927.
129. "Victory! Brazil Suspends New Industrial Logging in the Amazon as State," January 1, 1998, www.forests.org/archive/brazil/amazsusp.htm.
130. Nailene Chou, "Solution Sought as North's Forests Shrink along with Loggers Incomes," *South China Morning Post,* February 15, 2003.
131. Environment, Science, and Technology Section, U.S. Embassy Beijing, "The Environment at the 2000 NPC Plenary," March 2000, www.usembassy-china.org.cn/english/sandt/NPCenviro.htm.

132. Philip Pan, "China's Chopsticks Crusade: Drive Against Disposables Feeds Environmental Movement," *Washington Post*, February 6, 2001, p. A01.

133. Xiaoyang Jiao, "Fighting Against Illegal Lumbering," *China Daily*, December 13, 2000.

134. "China Makes Stable Progress in Forest Protection," Xinhua News Agency, May 20, 2001.

135. Vaclav Smil, *China's Environmental Crisis* (Armonk, N.Y.: M. E. Sharpe, 1993), 61.

136. *China Environment News*, December 14, 2001.

137. Calum MacLeod and Lijia MacLeod, "My-Grain Headache," *ChinaOnline*, April 23, 2001.

138. Melinda Liu, "Waiting for Rain: China Is Struggling to Cope with Its Worst Drought in Fifty Years," *Newsweek*, August 21, 2000, 44.

139. Ron Gluckman, "The Desert Storm," *Asiaweek*, October 13, 2000, www.asiaweek.com/asiaweek/magazine/2000/1013/is.china.html.

140. "Yunnan Lake Pollution Controls Detailed," Xinhua News Agency, January 15, 1999.

141. EPB official, interview with author's research associate, Kunming, June 2000.

142. "China's Water Crisis: Part 2: Pollution," www.websiteaboutchina .com/envi/environment_1_2.htm.

143. "Water Pollution Controls at Dianchi Lake Pay Off," Xinhua News Agency, November 11, 2002.

144. EPB officials, interviews with author's research associate, Kunming June 2000.

145. Jasper Becker, "Putrid Lake Proof Environmental Policies Have Failed to Hold Water," *South China Morning Post*, October 15, 2001, p. 8.

146. "China's South-to-North Water Diversion Project," Xinhua News Agency, August 14, 2003.

147. "Update on China's South-North Water Transfer Project," Report from Embassy Beijing, June 2003, www.usembassy-china. org.cn/sandt/SNWT-East-Route.htm.

148. Patrick Tyler, "Huge Water Projects Supply Beijing by 860-Mile Aqueduct," *New York Times*, July 19, 1994, p. A8.

149. "Update on China's South-North Water Transfer Project," June 2003.

150. "Chinese Premier: Water Diversion Project to Start in All-Round Way in East, Central China," Xinhua News Agency, August 14, 2003.

151. "Update on China's South-North Water Transfer Project," June 2003.

152. Carter Brandon and Ramash Ramankutty, "Toward an Environ-

mental Strategy for Asia," World Bank discussion group paper, Washington, D.C., 1993.

153. Chinese Academy of Social Sciences' official, interview with author, Beijing, March 8, 1995.

Chapter 5. The New Politics of the Environment

1. NGOs in China are not the same as their western counterparts. They must be sponsored by a government institution, which has nominal oversight for the NGOs' activities and membership. Still, they represent a substantial advance in independence from the government-organized NGOs that continue to predominate in Chinese society.

2. Minxin Pei, "Democratization in the Greater China Region," *Access Asia Review* 1, no. 2 (1998): 5–40.

3. Environment, Science, and Technology Section, U.S. Embassy in Beijing, "Environment Tops Urban Worries," *Beijing Environment, Science, and Technology Update,* November 3, 2000, www .usembassy-china.org.cn/english/sandt/estnews1103.htm.

4. David Hsieh, "China's Top 2 Concerns," *Straits Times* February 28, 2002.

5. "NGOs Develop Fast in China," *People's Daily* (English), November 10, 2002, www.english.peopledaily.com.cn/200211/10/eng20021110 _106567.shtml.

6. "Chinese NGOs—Carving a Niche Within Constraints," January 2003 report, U.S. Embassy in Beijing, www.usembassy-china.org.cn/ sand/ptr/ngos-prt.htm.

7. Anna Brettell, "Environmental NGOs in the People's Republic of China: Innocents in a Co-opted Environmental Movement?" *Journal of Pacific Asia* 6 (2000): 34.

8. Ibid., 35.

9. Falun Dafa is the spiritual movement that practices Falun Gong (Practice of the Wheel of the Dharma), a set of exercises and guiding principles that are said to lead its adherents to a higher dimension. The Chinese government, however, has labeled Falun Dafa a cult, and the practice of Falun Gong is forbidden in China.

10. Jian Song, speech at the Fourth National Conference on Environmental Protection, July 1996, in Qing Dai and Eduard B. Vermeer, "Do Good Work, But Do Not Offend the 'Old Communists,'" in *China's Economic Security,* ed. Robert Ash and Werner Draguhn (New York: St. Martins, 1999), 144.

11. Jiang Zemin, "Hold the Great Banner of Deng Xiaoping Theory for an All Around Advancement of the Cause of Building Socialism with Chinese Characteristics into the 21st Century," in *Beijing Review* 6–12 (October 1997): 10–33, cited in Tony Saich, "Negotiating the State: The Development of Social Organizations in China," *China Quarterly* 161 (March 2000): 128.

12. U.S. NSC staff member, conversation with author, June 1998.

13. State Council of the People's Republic of China, *Regulations for Registration and Management of Social Organizations*, People's Republic of China State Council Order No. 250, translated from *People's Daily*, April 11, 1998, in *China Development Brief*, www.chinadevelopmentbrief.com/page.asp?sec=2&sub=1&pg=1.

14. Elizabeth Knup, *Environmental NGOs in China: An Overview*, China Environment Series, no. 1 (Washington, D.C.: Woodrow Wilson Center Press, 1997), 10.

15. John Pomfret, "Chinese Crackdown Mixes Repression with Freedom," *Washington Post*, December 19, 1998, p. A14.

16. Jasper Becker, "Tightening the Noose on Parties," *South China Morning Post*, December 5, 1998, p. 2.

17. State Council, *Regulations for Registration and Management*.

18. Becker, "Tightening the Noose," p. 2

19. "China: Freedom of Association Regulated Away," www.three freedoms.org/finalreport2/1-cfoa.htm.

20. Saich, "Negotiating the State," 133.

21. Ibid., 126.

22. "Chinese NGOs—Carving a Niche Within Constraints."

23. U.S. Embassy official, interview with author, Beijing, March 2000.

24. Brettell, "Environmental NGOs in the PRC," 47.

25. Agenda 21 official, interview with author, Beijing, March 21, 2000.

26. Fengshi Wu, *New Partners or Old Brothers? GONGOs in Transnational Environmental Advocacy in China*, China Environment Series, no. 5 (Washington, D.C.: Woodrow Wilson Center Press, 2002), 56–57.

27. This discussion excludes the many research-oriented environmental NGOs. Their work is discussed in chapter 6, in the context of the international community's involvement in China's environmental situation.

28. Chinese environmental activist, interview with author, Beijing, July 6, 2000.

29. Ibid.

30. Tang Xiyang and Marcia Marks, *A Green World Tour* (New World Press: Beijing, 1999), insert for p. 240.

31. Ibid., insert for p. 221.
32. Ibid., insert for p. 240.
33. Chinese environmental activist, interview with author, Beijing, July 6, 2000.
34. Bochuan He, *China on the Edge* (San Francisco: China Books and Periodicals, 1991), 209.
35. Mark Hertsgaard, " 'What We Need to Survive City Dwellers Soon Will Outnumber Country Dwellers. But Can We Survive Our Mass Urbanization?' Previews from the Developing World Signal Trouble," *Boston Globe*, January 24, 1999, p. F1.
36. Bochuan He, *China on the Edge*, 43–49.
37. Ibid., 50–51.
38. Hertsgaard, "What We Need to Survive," p. F1.
39. Ai Wang, "Chinese Environmentalist Dai Quing [*sic*] Speaks Out on Three Gorges Dam," *Environment News Service*, May 26, 1999, www.threegorgesprobe.org/tgp/print.cfm?contentID=2808.
40. Ibid.
41. The English-language version of the book includes several essays not in the 1989 Chinese version, including a letter by Li Rui, Mao Zedong's secretary for industrial affairs and a vice minister of the Ministry of Water Resources and Electric Power, to Jiang Zemin urging that the project be canceled.
42. Qing Dai, *Yangtze! Yangtze!*, ed. Patricia Adams and John Thibodeau, trans. Nancy Liu, Mei Wu, Yougeng Sun, and Xiaogang Zhang (Toronto: Earthscan Publications, 1994), 136.
43. Ibid., xxiii.
44. Mark Levine, "And Old Views Shall Be Replaced By New," *Outside*, October 1997, www.outsidemag.com/magazine/1097/9710oldviews.html.
45. John Pomfret, "Dissidents Back China's WTO Entry: Trade Status Said Essential for Improved Human Rights," *Washington Post*, May 11, 2000, p. A01.
46. Chinese environmental activist, interview with author's research associate, Beijing, July 6, 2000.
47. In addition to the formally approved name, Liang wrote Friend of Nature (later changed to Friends) under the Chinese name on his seal. Friends of Nature then became the name by which the organization was known.
48. Chinese environmental activist, interview with author, Beijing, April 2000.
49. Dai and Vermeer, "Do Good Work," 147.

50. Chinese environmental activist, interview with author, Beijing, April 2000.
51. "About Friends of Nature," *Friends of Nature News* 2 (1999):8.
52. Chinese environmental activist, interview with author, New York, 1998.
53. "Activist Writer Wang Lixiong Dismissed from Environmental Group," iso.hrichina.org/iso/news.
54. George Wehrfritz, "Green Heat," *Newsweek,* October 7, 1996, 13.
55. Martin Williams, "The Year of the Monkey," *BBC Wildlife* (January 2000): 72.
56. Ibid., 73.
57. Chinese environmental activist, interview with author, New York, June 2000.
58. Williams, "Year of the Monkey," 74.
59. Chinese environmental activist, interview with author, Beijing, April 2000.
60. Wehrfritz, "Green Heat," 13.
61. Chinese environmental activist, interview with author, Beijing, April 2000.
62. Chinese environmental activist, interview with author, New York, June 2000.
63. The Chinese name, *Beijing Zhinong Shengtai Baohu Fazhan Yanjiu Zhongxin,* translates into Beijing Zhinong Research Center for Ecological Conservation and Development. This name includes Xi Zhinong's given name—Zhinong—and has remained the same even as the English name has changed.
64. Lihong Shi, co-founder, Green Plateau Institute, talk at Woodrow Wilson Center, Washington, D.C., December 8, 2000.
65. Ibid.
66. Lihong Shi, interview with author, New York, February 2002.
67. Guy Trebay, "Dead Chic," *Village Voice,* May 26–June 1, 1999.
68. "About Friends of Nature," 2.
69. Peter Popham, "These Animals Are Dying Out, and All Because the Lady Loves Shahtoosh," *The Independent,* June 20, 1998, reprinted in "About Friends of Nature," 7.
70. "Wild Yak Brigade Rides to the Rescue of the Rare Chiru," *U.S. News and World Report,* November 22, 1999, 38.
71. Congjie Liang, "Many Successes—But Much More Still Needs to Be Done," *Friends of Nature News* 2 (1999), www.fon.org.cn/newsletter/99–2e/4.html.

72. Chinese NGO leaders, interview with author, Woodrow Wilson Center, Washington, D.C., December 8, 2000.

73. About half of the NGO's revenues are supplied by the Hong Kong–based branch of Friends of the Earth; private companies in Shenzhen and Beijing also fund the NGO. Yang also receives support from the Canadian Civil Society Program and the WWF.

74. Chinese environmental activists, interview with author's research associate, Kunming, June 2000.

75. Differences over how to use the grant from CIDA have reportedly caused Tian and Wu to dissolve their partnership.

76. Chinese environmental activist, interview with author's research associate, Kunming, Yunnan, June 2000.

77. Changqun Duan and Xueqing Yang, "Political Ecology in Mainland China's Society for Fifty Years—Evolution of Responses of Political Affairs to Eco-Environmental Issues," draft paper on file with author, p. 10.

78. Chinese environmental activist, interview with author, New York, March 2000.

79. "Individuals Changing the World," *Beijing Review*, August 14, 2000, p. 20.

80. Ibid., 16.

81. Ibid., 17.

82. China State Environmental Protection Agency and Ministry of Education, *Quanguo Gongzhong Huangjing Yishi Diaocha Baogao* [Survey Report on Public Environmental Consciousness in China] (Zhongguo Huangjing Kexue Chubanshe, 1999), 21.

83. Ibid.

84. Wang Yongchen, interview with author, New York, February 2002; Kelly Haggart and Mu Lan, "People Power Sinks a Dam," *Three Gorges Probe News Service*, October 16, 2003.

85. Xiaowei Chen, "The Multiple Roles of Media in Today's Chinese Society" (paper presented at the "Memory and Media in and of Contemporary China" conference, University of California, Berkeley), in Xiaoping Li, *Significant Changes in the Chinese Television Industry and Their Impact in the PRC—An Insider's Perspective* (Washington, D.C.: Center for Northeast Asian Policy Studies, Brookings Institution, August 2001), 13, www.brookings.org/dybdocroot/fp/cnaps/papers/li_01.pdf.

86. Xiaoping Li, *Significant Changes*, 8.

87. Ibid., 10.

88. *China Environment News,* May 15, 1995, p. 1.
89. Chinese environmental activist, interview with author, Beijing, April 2000.
90. Ying Wang, "Re-energizing Battery Recycling Efforts," *China Daily,* August 7, 2000.
91. Jun Liu and Qian Zeng, "Public Take Recycling into Their Own Hands," *China Daily,* September 8, 2000.
92. Ibid.
93. For a discussion of such activities, see Dai and Vermeer, "Do Good Work," 154.
94. Kanping Hu (editor, *China Green Times*), talk at Woodrow Wilson Center, Washington, D.C., December 8, 2000.
95. Bo Wen, *Greening the Chinese Media,* China Environment Series, no. 2 (Washington, D. C.: Woodrow Wilson Center Press, 1998), 39.
96. Ibid., 39.
97. Chinese environmental activist, interview with author, Dalian, April 2000.
98. Environmental activist, talk in Washington, D.C., December 8, 2000.
99. Chinese environmental activist, interview with author, Beijing, June 2002.
100. The island of Hong Kong was actually leased to Britain in the 1890s, and the lease expired in 1997. A small area on the mainland that had been under British control since the end of the Opium War in the 1840s was also returned by the British to the mainland in 1997.
101. For a detailed discussion of NGOs in Hong Kong, see Stephen Wing-Kai Chiu, Ho-Fung Hung, and On-Kwok Lai, "Environmental Movements in Hong Kong," in *Asia's Environmental Movements,* ed. Yok-shiu F. Lee and Alvin Y. So (Armonk, N.Y.: M. E. Sharpe, 1999), 55–89.
102. Ibid., 62–72.
103. Jennifer Ehrlich, "Anger over Plan for Third Nuclear Plant," *South China Morning Post,* January 19, 2000, p. 1.
104. Ibid.
105. Ng Cho Nam (director, Conservancy Association, Hong Kong), talk at the Woodrow Wilson Center Working Group on Environment in U.S.–China Relations, Washington, D.C., December 6, 2000.
106. Antoine So, "Don't Talk to Me about Tree Frogs," *South China Morning Post,* October 6, 2001, p. 11.
107. Wen, *Greening the Chinese Media,* 42.
108. Jennifer Ehrlich, "Cadres Prepared to Hear Concerns," *South China Morning Post,* May 3, 2000, p. 8.

109. "Environmental Groups Join Beijing's Bid for Green Olympics," *People's Daily*, August 25, 2000, fpeng.peopledaily.com.cn/200008/24/eng20000824_48957.html.

110. "Beijing Relocates Steel-Making Capacity to Hebei Province," New China News Agency, January 20, 2003, as reported in BBC Monitoring International Reports.

111. Dai and Vermeer, "Do Good Work," 147.

112. Amanda Bower, "Not Everyone in China Is Cheering," *Time*, July 23, 2001, 10.

Chapter 6. The Devil at the Doorstep

1. This is discussed at length in chapter 4.

2. For an in-depth examination of China's process of accession to international environmental treaties and its record on implementation of those treaties, see Michel Oksenberg and Elizabeth Economy, "China's Accession to and Implementation of International Environmental Accords 1978–1995," occasional paper, Asia/Pacific Research Center, Stanford University, February 1998.

3. Ibid., 13.

4. For the extraordinary story of World Wildlife Fund's first, tumultuous cooperative venture in China, see George Schaller, *The Last Panda* (Chicago: University of Chicago Press, 1993).

5. Oksenberg and Economy, "China's Accession," 29–30.

6. Elizabeth Economy, "Negotiating the Terrain of Global Climate Policy in the Soviet Union and China: Linking International and Domestic Decisionmaking Pathways" (Ph.D. diss., University of Michigan, 1994), 159.

7. Ibid., 179.

8. State Science and Technology Commission officials, interviews with author, Beijing, June 1994 and May 1996.

9. State Planning Commission official, interview with author, Beijing, June 1992.

10. *Joint implementation* was a scheme proposed by the advanced industrialized countries whereby they would receive credit toward achieving their targets for greenhouse gas reduction by undertaking projects in developing countries (e.g., a U.S. power company could pursue a reforestation effort in Brazil) where the cost of taking action would be substantially lower than the domestic cost.

11. The rules governing CDM have yet to be fully worked out.

12. Several U.S.–based research organizations, including Lawrence Berkeley Lab, World Resources Institute, and the Natural Resources Defense Council have reported that their calculations support such a claim of reduced emissions; other experts, however, remain unconvinced by the analysis. John Pomfret, "Research Cast Doubt on China's Pollution Claims," *Washington Post*, August 15, 2001, p. A16.

13. Martin Lees, "China and the World in the Nineties" (conference summary report, Beijing, January 25, 1991), 30–31.

14. Ibid., 9–19.

15. Ibid., 22.

16. Foreign Broadcast Information Service [hereafter FBIS], China Daily Report, December 27, 1990, p. 31.

17. The sponsoring entities for CCICED later expanded to include the United Kingdom, Germany, the Netherlands, and the World Bank in the wake of the UNCED. The CCICED has anywhere from five to seven short-term task forces focusing on issues such as resource accounting and pricing, biodiversity, trade and environment, and energy. It consists of fifty Chinese and foreign environment and economics experts, as well as Chinese officials and scientists. The group meets annually and issues reports that provide concrete recommendations for the government. November 2002 recommendations included enhancing the role of private organizations in environmental enforcement and employing eco-taxes. Recommendations are often directly delivered to top-level Chinese officials because some Chinese participants themselves are vice ministers or even ministers. (They also run seminars and workshops to disseminate key information.)

18. These included protection of sovereignty over natural resources, free transfer of technology from the more to the less advanced industrialized countries, and the responsibility of the advanced industrialized countries to bear the cost of addressing the global environmental problems to which it contributed so heavily.

19. See chapter 3 for a full discussion of UNCED's impact on Chinese domestic environmental practices.

20. Correspondence with U.S.–based NGO leader on file with author, November 2001.

21. FBIS, China Daily Report, June 29, 1994, p. 22.

22. FBIS, China Daily Report, July 11, 1994, p. 31. Officially, in 1996, the Chinese claimed that one-third of the $4 billion necessary to fund these projects had been raised. One SSTC official, however, suggested that this figure was likely inflated.

23. Not all Chinese officials welcome the deeper integration of international environmental protection values, attitudes, and approaches

with those of China. Despite now widespread acceptance of the ideals of sustainable development, for example, the term's western roots became problematic in 1995–1997. There was increasing discussion in the Chinese media suggesting that sustainable development was part of a master plan by advanced industrialized countries (and especially the United States) to contain China by forcing it to slow the pace of economic growth in order to protect the environment. While this period appears to have been a temporary aberration in an overall trend toward accepting the ideals embraced by the concept of sustainable development, it suggests that environmental protection, like many other multilateral issues, has the potential to be easily politicized.

24. Agenda 21 Center official, interview with author, Beijing, April 2000.

25. Notably, SEPA is the secretariat for CCICED and co-sponsor of the World Bank report.

26. Daniel Esty and Seth Dunn, "Greening U.S. Aid to China," *China Business Review* (January–February 1997): 41–45.

27. The Ministry of Foreign Affairs of Japan, "Japan's Environmental ODA for China," www.mofa.go.jp/policy/oda/category/environment/pamph/2001/china.html.

28. Correspondence with U.S.–based NGO official on file with author, November 2001.

29. Xueqing You, "The Blue Skies of Clean Beijing," *Science and Technology Daily*, January 11, 2000, 2 (in Chinese); available in English in Environment, Science, and Technology Section, U.S. Embassy Beijing, "Notes: Beijing Air Quality Improves" (October 2000), www.usembassy-china.org.cn/sandt/bjairup.html.

30. NRDC official, phone interview with author, February 2002.

31. "ADB Urges China to Take Steps to Protect Environment," *Asia Pulse*, October 16, 2001.

32. China Council for International Cooperation on Environment and Development [CCICED], "CCICED Recommendations to the Chinese Government," www.harbour.sfu.ca/dlam/recommendations/.

33. Shanghai EPB official, interview with author, Shanghai, September 1999.

34. "Water Crisis, Part I," www.websiteasaboutchina.com/envi/environment_1_1.htm.

35. William P. Alford and Yuan Yuan Shu, "Limits of the Law in Addressing China's Environmental Dilemma," *Stanford Environmental Law Journal* (January 1997): 136.

36. Aminul Huq, Bindu N. Lohani, Kazi F. Jalal, and Ely A. R. Ouano, "The Asian Development Bank's Role in Promoting Cleaner Produc-

tion in the People's Republic of China," *Environmental Impact Assessment Review* 19 (1999): 550–51.

37. Huq et al., "Asian Development Bank's Role," 542.

38. Some corporations, such as BP, have moved well beyond supporting only capacity building directly related to their immediate needs and are also focusing on China's long-term environmental outlook. BP has been working for seven years with the World Wide Fund for Nature to support the inclusion of environmental education in a range of subjects at three teaching universities. After a three-year trial run, the program is now being expanded to ten universities, and developing certificate and masters degree courses in environmental education. On a much smaller scale, Shell has supported renewable energy projects such as the development of biogas in Yunnan and gasified straw in Zhejiang.("Multinationals Meet to Explore 'Corporate Social Responsibility in China,'" *China Development Brief* 4, no. 1 (2001).

39. Kathy Wilhelm, "Still Spewing, Just a Little Less," *Far Eastern Economic Review,* July 5, 2001, 34.

40. The international demand for partnership has far outpaced the capabilities of the limited number of qualified Chinese counterparts. Venturing outside the major cities and well-known Chinese institutions greatly increases the start-up costs for the international partner. The top Chinese research institutes, universities, and think tanks, however, are full, if not over-subscribed, with international cooperative ventures. This produces its own challenges. In one case, for example, a U.S.–based think tank expended several years and almost $120,000 in travel and research expenses before its prominent, well-respected Chinese partner admitted that he would not be able to undertake the research project he had promised because his time was occupied with several other substantial international projects.

41. Director, Beijing-based U.K. multinational, e-mail correspondence on file with author, March 2002.

42. Huq et al., "Asian Development Bank's Role," 551.

43. *Cleaner production* is a "conceptual and procedural approach to production that demands that all phases of the lifecycle of a product or a process be addressed with the objectives of prevention and minimization of short- and long-term risks to human health and to the environment" (Huq et al., "Asian Development Bank's Role," 543).

44. Ibid.

45. World Bank, *China: Air, Land, and Water* (Washington, D.C.: World Bank, 2001), 2.

46. Cleaner production is also a central element of China's Agenda 21 action plan.

47. Huq et al., "Asian Development Bank's Role," 546–49.
48. GM officials, phone interview with author, December 2000.
49. Charles W. Schmidt, "Economy and Environment: China Seeks a Balance," *Environmental Health Perspectives* 110, no. 9 (September 2002), 516–22.
50. Nagle, "Missing Environmental Law," 537.
51. Ibid.
52. World Bank, *China: Air, Land, and Water*, 104.
53. Jim Watson, Xue Liu, Geoffrey Oldham, Gordon Mackerron, and Steve Thomas, "International Perspectives on Clean Coal Technology Transfer to China: Final Report to the Working Group on Trade and Environment, China Council for International Cooperation on Environment and Development" (August 2000), 15.
54. The company's flagship Changsha power plant in Hunan Province, which is a Build-Operate-Transfer power plant, will be closely supervised and monitored by international lending institutions and SEPA. (The World Bank has explicitly used its financial leverage to press for the use of cleaner coal technologies in some of the power plants it has financed.) Their Jiaxing plant, in contrast, will be primarily owned and operated by a provincial power bureau. National power is concerned about Jiaxing's reluctance to adhere to SEPA regulations and its willingness simply to pay a nominal pollution fee to avoid spending money on abatement equipment. (Watson et al., "International Perspectives," 26.)
55. The case of the U.S. firm, Combustion Engineering, has become part of the lore of multinationals' experience in China. After Combustion Engineering licensed its designs to the Ministry of Electric Power, the Ministry shared the designs with all of China's large boiler makers, failing to compensate Combustion Engineering appropriately in the process. Even the Global Environmental Facility, whose Chinese counterpart is the powerful Ministry of Finance, has encountered serious problems in this regard. A large-scale project to "subsidise the acquisition of technology licenses for new industrial boiler technologies by Chinese firms" failed to attract top multinationals in part because of their concern that their intellectual property would not be respected by their Chinese partner (Watson et al., "International Perspectives," 55). The experience of Combustion Engineering has chastened other companies, such as Mitsui Babcock, which has resisted licensing its technology. Some multinationals attempt to protect themselves by identifying some parts of a technology or process that they are willing to transfer and some that they are not. Shell, for example, transfers the know-how embodied in its gasification process,

but retains the design of key components to preserve the company's technological and market advantage. Others, such as General Motors, may not only share the technology but also work jointly with Chinese partners on basic R&D, investing Chinese managers, scientists, and experts in the long-term future of the joint venture. GM's venture with Shanghai Automotive, for example, includes basic research and collaboration on engineering, software, and biomechanics, among other areas of technology development. Chinese managers and technicians are also sent to GM plants in the rest of the world, and GM places foreign managers and technicians in its Chinese venture. Still, GM does not share some of its most advanced technology, for example, with regard to fuel cells, for national security reasons.

56. Watson et al., "International Perspectives," 36.

57. Ibid., 29.

58. A joint effort among the Massachusetts Institute of Technology, Qinghua University, and the Swiss Federal Institutes of Technology to improve boiler efficiency in 256 sites in Henan, Jiangsu, and Shanxi Provinces initially failed, despite developing a number of well-tailored and inexpensive measures for the various boiler sites. The MIT-led team discovered that the SOE directors had no immediate incentive to upgrade the technologies in their plants. The team is now researching the linkage between pollution and local health costs to help persuade local officials and enterprise directors of the importance of the project as well as searching out mechanisms for funding to implement the upgrades. (At the same time, the team is confronting resistance from local public health officials, who do not want to share information.) Watson et al., "International Perspectives," 50.

59. World Bank, *China: Air, Land, and Water*, 102–4.

60. Terence Tsai, Stefan Eghbalian, and Hans Tsai, "China's Green Challenge," *Harvard China Review* 2, no. 1 (2000): 85.

61. Ibid., 85.

62. Ibid., 85.

63. World Bank, *China: Air, Land, and Water*, xxii.

64. Peter Morici, *Reconciling Trade and the Environment in the World Trade Organization* (Washington, D.C.: Economic Strategy Institute, 2002), 8.

65. Changhua Wu, "Trade and Sustainability—A China Perspective," *Sinosphere* 3, no. 3 (2000): 29.

66. China Human Development Report, "2002: Making Green Development a Choice" (UN Development Programme and Stockholm Environment Institute), 63.

67. Wu, "Trade and Sustainability," 28–33.

68. Tao Hu and Wanhua Yang, "Environmental and Trade Implications of China's WTO Accession" (unpublished paper [on file with author] prepared for the Working Group on Trade and Environment CCICED, September 2000), 14.

69. Michael Dorgan, "China's Food Hazards Take on Global Significance," *The Seattle Times,* March 17, 2002, p. A16.

70. Hu and Yang, "Environmental and Trade Implications," 5–8.

71. Ibid., 6.

72. Ibid., 5–8.

73. Asia-Pacific Economic Cooperation [APEC], "Summary: APEC in Sustainable Development 1999," http://203.127.220.67/apec_groups/other_apec_groups/sustainable_development.download links.0002.linkURL.download.ver5.1.9.

74. Asia-Pacific Economic Cooperation [APEC], "APEC Energy Ministers Joint Statement on Clean Energy and Sustainable Development" (May 12, 2000), http://203.127.220.67/apec/ministerial_statements/sectoral_ministerial/energy/00energy/00cleanenergy.html.

75. "China Vows to Protect Sea from Pollution," *People's Daily,* October 10, 2000, fpeng.peopledaily.com.cn/200010/10/eng20001010_52194.htm.

76. "Summary: APEC in Sustainable Development 1999."

77. "China Says APEC Shanghai Summit Is a Success," Xinhua News Agency, October 21, 2001.

78. Michael Richardson, "Sharing the Mekong: An Asia Challenge," *International Herald Tribune,* October 30, 2002, p. 2.

79. For an informative and insightful look at the historical and contemporary issues surrounding the Three Gorges Dam, see Qing Dai, *Yangtze! Yangtze!,* ed. Patricia Adams and John Thibodeau, trans. Nancy Liu, We Mei, Sun Yongeng, Zhang Xiaogang (London: Earthscan, 1994).

80. Current director of the World Bank James Wolfenson has increased the range of "safeguards" accounted for in any project assessment. Gender, resettlement, minority populations, environment, and public participation are all factors that must be considered before the Bank will fund a project. In China, such considerations have hampered several potential collaborations.

81. Karen Cook, "Dam Shame," *Village Voice,* March 29–April 4, 2000, p. 48, www.villagevoice.com/issues/0013/cook.php.

82. "Three Gorges Work 'Shoddy,'" *Financial Times,* June 8, 1999, www.irn.org/programs/threeg/990608.ft.html410A.

83. Kelly Haggart, "SARS and Falun Gong Provide Pretexts for Three Gorges Arrests," *Three Gorges Probe News Service,* August 14, 2003, www.threegorgesprobe.org/tgp/index.cfm?DSP=Content&ContentID=8122.

84. Kelly Haggart, "More Cash Needed to Fix 'Enormous' Resettle-

ment Problems, Official Says," March 14, 2003, www.threegorges
probe.org/tgp/index.cfm?DSP=content&ContentID=6784.

85. John Pomfret, "China's Giant Dam Faces Huge Problems," *Washington Post,* January 7, 2001, p. A1.

86. "Chinese Officials Alarmed at Looming Environmental Crisis at Three Gorges Dam," *Three Gorges Probe News Service,* February 14, 2001, ww.threegorgesprobe.org/tgp/index.cfm?DSP=content&ContentID=1718.

87. Ray Cheung, "Three Gorges Cleanup Is Faltering, Says Official," *South China Morning Post,* March 5, 2003.

88. "Water Quality in Dam Zone Maintained," *China Daily,* June 6, 2003, www.crienglish.com/144/2003-6-6/19@17850.htm.

89. Luke Harding, "BP Caught in Tibet Crossfire," *The Guardian,* October 2, 2000, p. 15.

90. "Top Chinese Leaders Discuss Policy, Development of Western Regions," Xinhua News Agency Domestic Service (in Chinese), January 23, 2000, translated in BBC Worldwide Monitoring, January 24, 2000.

91. "China's Premier Invites Foreigners to Invest," *Asia Pulse,* March 16, 2000.

92. Andrew Browne, "China Looks West to Underpin Prosperity," Reuters English News Service, March 6, 2000.

93. "China Basic Statistics by Province," www.ers.usda.gov/Briefing/China/Data/tbl-provinces.xls.

94. William Kazer, "Poor Regions in Line for US$1b World Bank Aid," *South China Morning Post,* March 18, 2000, p. 3; Ted Plafker, "China's Go West Drive," *International Herald Tribune,* May 8, 2001, p. 9.

95. Mary Kwang, "Inland Regions to Get Funds First," *Straits Times,* March 7, 2000.

96. "Economist Calls for New Strategies in Developing Western China," Zhongguo Tongxun She News Agency, in BBC Worldwide Monitoring, January 6, 2000.

97. Ibid.

98. "China: Checking Pollution," *China Daily,* January 10, 2000, p. 1.

99. "PRC to Build Nature Reserves on Qinghai-Tibet Plateau," *World News Connection,* March 9, 2000.

100. Environment, Science, and Technology Section, U.S. Embassy Beijing, "Environment at the 2000 NPC Plenary," www.usembassy-china.org.cn/sandt/NPCenviro.htm.

101. "Xuezhe Weiyuan Retan Xibu" [Scholars from the People's Political Consultative Conference enthusiastically discuss the western

regions], March 7, 2000, *Zhongugo Huanjing Bao* [China Environment News], www.ee65.com.cn.

102. Tao Fu, "Chinese Scholars Call for Privatization in the West," *China Development Brief* 4, no. 1 (2001), www.chinadevelopment brief.com/article.asp?sec=19&sub=1&toc=1&art=356.

103. "China to Learn from Western Countries in 'Go-West' Campaign," *People's Daily*, March 8, 2003, www.english.peopledaily .com.cn/200303/08/eng20030308_112970.shtml.

104. Allen T. Cheng, "Western Province Projects Get $122 billion Injection," *South China Morning Post*, March 26, 2003, p. 8.

105. The Han are the dominant ethnic group in the country equaling 91.59 percent of the population. (National Bureau of Statistics of China, *Di Wu Ci Quanguo Renkou Pucha Gongbao [Diyi Hao]* [Report from the Fifth National Census (no. 1)], May 15, 2001, www.stats.gov.cn/tjgb/rkpcgb/qgrkpcgb/200203310083.htm.)

106. Browne, "China Looks West."

107. "Top Chinese Leaders Discuss Policy, Development of Western Regions," Xinhua News Agency Domestic Service (in Chinese), January 23, 2000, translated in BBC Worldwide Monitoring, January 24, 2000.

108. Plafker, "China's Go West Drive," 9.

109. "Hu Jintao on Development of West, Tibet," Xinhua News Agency, March 5, 2000.

110. Michael M. Phillips, "World Bank Delays a Plan on Resettlement," *Asian Wall Street Journal*, June 25, 1999, p. 2.

111. Alison Reynolds, "Tibetan Takeover," *The Ecologist* 30, no. 6 (2000): 20.

112. While the international community, and the United States in particular, may be sympathetic to China's fears of terrorism in Xinjiang, President Bush has clearly signaled that the United States will differentiate between ethnic separatist movements and terrorism.

113. Justin Lowe "The Scorched Earth: China's Assault on Tibet's Environment," *Multinational Monitor* (October 1992), multinational-monitor.org/hyper/issues/1992/10/mm1092.html#scorched.

114. "Fueling the Economic Miracle," *South China Morning Post*, March 22, 2003, p. 11.

Chapter 7. Lessons from Abroad

1. Susan Baker and Bernd Baumgartl, "Bulgaria: Managing the Environment in an Unstable Transition," in *Dilemmas of Transition: The*

Environment, Democracy, and Economic Reform in East Central Europe, ed. Susan Baker and Petr Jehlicka, special issue, *Environmental Politics* 7, no. 1 (1998): 189.

2. Fred Singleton, "Czechoslovakia: Greens versus Reds," in *Environmental Problems in the Soviet Union and Eastern Europe*, ed. Fred Singleton (Boulder, Colo.: Lynne Rienner Publishers, 1987), 175–76.

3. Singleton, "Czechoslovakia: Greens versus Reds," 176.

4. Joan DeBardeleben, "The Future Has Already Begun: Environmental Damage and Protection in the GDR," in *To Breathe Free: Eastern Europe's Environmental Crisis*, ed. Joan DeBardeleben (Washington, D.C.: Woodrow Wilson Center Press, 1991), 179.

5. Ibid., 177–78.

6. Joanne Landy and Brian Morton, "Perestroika May Be Both Good and Bad for Eastern Europe's Severe Ecological Crisis," *Utne Reader*, January/February 1989, 86.

7. Miklós Persányi, "Social Support for Environmental Protection in Hungary," in DeBardeleben, *To Breathe Free*, 213–14.

8. Michael Waller, "Geopolitics and the Environment in Eastern Studies," *Environmental Politics* 7, no. 1 (1998): 32.

9. Frances Millard, "Environmental Policy in Poland," *Environmental Politics*, 7, no. 1 (1998): 145.

10. In some cases the functions of environmental protection were merely grafted on to previously existing ministries of agriculture, forestry, or water management. In others, such as Poland, the GDR, and Hungary, in the mid-1980s, environmental protection agencies were granted ministerial status (Barbara Jancar-Webster, "Environmental Politics in Eastern Europe in the 1980s," in DeBardeleben, *To Breathe Free*, 29).

11. DeBardeleben, "The Future Has Already Begun," 185.

12. Susan Baker and Peter Jehlicka, "Dilemmas of Transition: the Environment, Democracy, and Economic Reform in East Central Europe—An Introduction," *Environmental Politics*, 7, no. 1 (1998): 8.

13. Jancar-Webster, "Environmental Politics in Eastern Europe, 25.

14. Michael Waller and Frances Millard, "Environmental Politics in Eastern Europe," *Environmental Politics* 1, no. 2 (1992): 164.

15. Singleton, "Czechoslovakia: Greens versus Reds," 180.

16. Waller and Millard, "Environmental Politics in Eastern Europe," 164.

17. Frances Millard, "Environmental Policy in Poland," *Environmental Politics* 7, no. 1 (1998): 146.

18. Jancar-Webster, "Environmental Politics in Eastern Europe," 38.

19. Singleton, "Czechoslovakia: Greens versus Reds," 180.
20. Christine Zvosec, "Environmental Deterioration: Eastern Europe," *Survey* 28, no. 4 (1984): 135.
21. DeBardeleben, "The Future Has Already Begun," 176.
22. Baker and Jehlicka, "Dilemmas of Transition," 9.
23. Persányi, "Social Support for Environmental Protection in Hungary," 216.
24. Zvosec, "Environmental Deterioration," 138–39.
25. Jancar-Webster, "Environmental Politics in Eastern Europe," 40.
26. Zvosec, "Environmental Deterioration," 136.
27. Millard, "Environmental Policy in Poland," 147–48.
28. Zvosec, "Environmental Deterioration," 136.
29. Persányi, "Social Support for Environmental Protection in Hungary," 218.
30. DeBardeleben, "The Future Has Already Begun," 176.
31. Zvosec, "Environmental Deterioration," 137.
32. Jancar-Webster, "Environmental Politics in Eastern Europe," 44.
33. Robert Peter Gale and Thomas Hauser, *Final Warning: The Legacy of Chernobyl* (New York: Warner Books, 1988), 27, cited in Murray Feshbach and Alfred Friendly Jr., *Ecocide in the USSR* (New York: Basic Books, 1992), 12.
34. Zhores Medvedev, *The Legacy of Chernobyl* (New York: W.W. Norton, 1990), 57, quoted in Feshbach and Friendly, *Ecocide in the USSR.*
35. Feshbach and Friendly, *Ecocide in the USSR,* 83
36. D. J. Peterson, *Troubled Lands: The Legacy of Soviet Environmental Destruction* (Boulder, Colo.: Westview Press, 1993), 200.
37. Ibid., 197.
38. Feshbach and Friendly, *Ecocide in the USSR,* 22.
39. Jancar-Webster, "Environmental Politics in Eastern Europe," 44.
40. Zvosec, "Environmental Deterioration," 139.
41. Ibid.
42. Jancar-Webster, "Environmental Politics in Eastern Europe," 41.
43. Baker and Jehlicka, "Dilemmas of Transition," 9.
44. Feshbach and Friendly, *Ecocide in the USSR,* 232.
45. Ibid., 223.
46. Ibid., 22.
47. Ibid., 232.
48. Jeffrey Glueck, "Subversive Environmentalism: Green Protest and the Development of Democratic Oppositions in Eastern Europe, Latvia, the Czech Lands, and Slovakia" (B.A. honors essay, Harvard University, 1991), 23–24.

49. Amanda Sebestyen, "Balkan Utopias," *Catalyst* (November 1990/January 1991), 28, in Waller and Millard, "Environmental Politics in Eastern Europe," 161.
50. Glueck, "Subversive Environmentalism," 26.
51. Ibid., 23.
52. Singleton, "Czechoslovakia: Greens versus Reds," 175.
53. Ibid., 179.
54. Jancar-Webster, "Environmental Politics in Eastern Europe," 43.
55. Zvosec, "Environmental Deterioration," 136.
56. Liliana Botcheva, "Focus and Effectiveness of Environmental Activism in Eastern Europe: A Comparative Study of Environmental Movements," *Journal of Environment and Development* (September 1996): 295–96.
57. Jancar-Webster, "Environmental Politics in Eastern Europe," 44–45.
58. Persányi, "Social Support for Environmental Protection in Hungary," 218–19.
59. Helsinki Watch Committee, "Independent Peace and Environmental Movements in Eastern Europe" (Helsinki Watch report, 1992), in Botcheva, "Focus and Effectiveness," 296–97.
60. Jancar-Webster, "Environmental Politics in Eastern Europe," 45–46.
61. Ronnie D. Lipschutz, "Damming Troubled Waters: Conflict over the Danube, 1950–2000," *Intermarium* 1, no. 2 (1997): 7, www.columbia.edu/cu/sipa/REGIONAL/ECE/dam.html.
62. Botcheva, "Focus and Effectiveness," 297.
63. Jancar-Webster, "Environmental Politics in Eastern Europe," 48.
64. Waller and Millard, "Environmental Politics in Eastern Europe," 165–66.
65. Botcheva, "Focus and Effectiveness," 299–300.
66. Peterson, *Troubled Lands,* 216.
67. Ibid., 216–17.
68. Ibid., 216.
69. Baker and Jehlicka, "Dilemmas of Transition," 10.
70. C. Thomassen, "Romania's Black Town," *World Press Review* 40, no. 9 (1993): 43, in Botcheva, "Focus and Effectiveness," 304.
71. Ibid.
72. Ibid., 304–5.
73. Peter Havlicek, "The Czech Republic: First Steps toward a Cleaner Future," *Environment* 39 (April 1997): 18.
74. "Environmental Trends in Transition Economies," Organisation of Economic Co-operation and Development (OECD) Policy Brief, *OECD Observer* (October 1999): 3.
75. Gusztav Kosztolanyi, "Where There's Much There's Brass," *Central*

Europe Review 1, no. 12 (September 13, 1999), www.ce-review.org/99/12/csardas12.html.

76. Gerald Fancoj, ed., *The Emerging Environmental Market: A Survey in Bulgaria, Croatia, Romania, and Slovenia* (Szentendre, Hungary: Regional and Environmental Center for Central and Eastern Europe, September 1997), 24, 49–50, www.rec.org/REC/publications/EmEnv Market2/EmEnvMarket2.pdf.

77. "Europe: Clean Up or Clear Out," *The Economist,* December 11, 1999, 47.

78. Laura A. Henry, "Two Paths to a Greener Future: Environmentalism and Civil Society Development in Russia," *Democratizatsiya,* spring 2002, 184–206.

79. Paul Brown, "US Backs Plan for Russia to Import Nuclear Waste," *The Guardian,* February 19, 2001, in Henry, "Two Paths to a Greener Future," 205.

80. Francisco A. Magno, "Environmental Movements in the Philippines," in *Asia's Environmental Movements,* ed. Alvin Y. So and Yok-shiu F. Lee (Armonk, N.Y.: M. E. Sharpe, 1999), 148.

81. Carter Brandon and Ramesh Ramankutty, *Toward an Environmental Strategy for Asia* (Washington, D.C.: International Bank for Reconstruction and Development, 1993), 21.

82. James Rush, *The Last Tree* (New York: The Asia Society, 1991), 4.

83. Chao-Chan Cheng, "A Comparative Study of the Formation and Development of Air and Water Pollution Control Laws in Taiwan and Japan," *Pacific Rim Law and Policy Journal* 3 (1993): S62.

84. *Newsreview,* May 21, 1994, p. 7.

85. Ubonrat Siriyuvasak, "The Environment and Popular Culture in Thailand," *Thai Development Newsletter* 26 (1994): 64–69, in Alvin Y. So and Yok-shiu F. Lee, "Environmental Movements in Thailand," in *Asia's Environmental Movements,* 120–21.

86. Carter Brandon, "Confronting the Growing Problem of Pollution in Asia," *Journal of Social, Political, and Economic Studies* 21 (summer 1996): 201.

87. Magno, "Environmental Movements in the Philippines," 148.

88. Brandon, "Confronting the Growing Problem of Pollution," 199.

89. So and Lee, "Environmental Movements in Thailand," 121.

90. Magno, "Environmental Movements in the Philippines," 146.

91. Michael Vatikiotis, "Malaysian Forests: Clearcut Mandate," *Far Eastern Economic Review,* October 28, 1993, 54.

92. "Was Illegal Logging Responsible for the Flash Flood in North Sumatra?" *Indonesian Mediawatch,* November 7, 2003, www.rsi.com.sg/english/indonesiamediawatch/view/2003110718209/1/.html.

93. Magno, "Environmental Movements in the Philippines," 149.

94. Lee and So, "Environmental Movements in Thailand," 140.

95. Hidefumi Imura, "Japan's Environmental Balancing Act: Accommodating Sustained Development," *Asian Survey* 34, no. 4 (1994): 356–57.

96. *Newsreview,* May 21, 1994, p. 7.

97. "Environmental Moves on Right Path," *Yomiuri Shimbun,* October 22, 2000.

98. Ibid.

99. "Minebea Co., LTD." www.minebea.co.jp/english/company/business/environment/history/history.html.

100. Colin MacAndrews, "Politics of the Environment in Indonesia," *Asian Survey* 34, no. 4 (1994): 377.

101. Rush, *The Last Tree,* 87.

102. Frances Seymour (director, Development Assistance Policy, World Wildlife Fund), phone conversation with author, on file with author, March 1996.

103. "NGOs to Call for Aid Boycott If Forest Law Not Enforced," *Jakarta Post,* February 16, 2001.

104. Azlan Abu Bakar, "Timber Certification Scheme Falls Victim to Lobbyists," *Business Times,* February 6, 2002, p. 2.

105. Chelsea L. Y. Ng, "When Courts Can't Do Their Bit for the Environs," *The Star,* January 27, 2002, www.wwfmalaysia.org/Newsroom/localnews/localnews.asp?selnews=1019&semonth=1&selyear=2002.

106. Lee and So, "Environmental Movements in Thailand," 123–24.

107. Ibid., 124–125.

108. *1992 World Development Report* (New York: Oxford University Press, 1992), 88.

109. Magno, "Environmental Movements in the Philippines," 150.

110. Ibid., 152.

111. Takamine Tsukasa, "Asia's Environmental NGOs: Emerging Powers?" *Jakarta Post,* June 14, 1999.

112. Su-Hoon Lee, Hsin-Huang Michael Hsiao, Hua-Jen Liu, On-Kwok Lai, Francisco A. Magno, and Alvin Y. So, "The Impact of Democratization on Environmental Movements," in *Environmental Movements in Asia,* ed. Lee and So, 233.

113. Ibid., 233.

114. Ibid., 234.

115. "Green Korea United Launches Legal Campaign to Save Environment," *Korea Times,* March 31, 2000.

116. Ibid.

117. "Environmental NGOs Have Lost Their Soul," *Nation*, May 18, 2000.
118. Ibid.
119. Hsin-Huang Michael Hsiao, On-Kwok Lai, Hwa-Jen Liu, Francisco A. Magno, Lara Edles, and Alvin Y. So, "Culture and Asian Styles of Environmental Movements," in *Environmental Movements in Asia*, ed. Lee and So, 215.
120. Ibid., 214.
121. Robert Mason, "Whither Japan's Environmental Movement? An Assessment of Problems and Prospects at the National Level," *Pacific Affairs*, July 1, 1999, 193.
122. Ibid., 197.
123. Ibid., 199.
124. Ibid., 198.
125. Lee and So, "Environmental Movements in Thailand," 128–29.
126. Ibid., 139.
127. Takamine Tsukasa, "Asia's Environmental NGOs: Emerging Powers?" *Jakarta Post*, June 14, 1999.
128. Mason, "Whither Japan's Environmental Movement?" 196.
129. Glueck, "Subversive Environmentalism," 30.
130. Feshbach and Friendly, *Ecocide in the USSR*, 232.
131. Ibid., 233.
132. Glueck, "Subversive Environmentalism," 2.
133. Ibid., 6.
134. Ibid., 22.
135. Ibid., 2–3.
136. Ibid., 28.
137. Eric Eckholm, "Spreading Protests by China's Farmers Meet with Violence," *New York Times*, February 1, 1999, p. A1; "Farmers Arrested: Police Arrest Nine Farmers Suspected of Leading a Rural Tax Protest," *South China Morning Post*, February 1, 1999, p. 8.
138. "Rights Group Claims China Sent Agents to Kidnap Escaped Activist," *AFP*, November 5, 2000.
139. "Brief Introduction of China Development Union," fdforum.org/cdu/cdujje.htm.
140. John Pomfret, "Reform Hot Topic of Group in Beijing; Open Debate Grows as Jiang Eases Grip," *Washington Post*, September 13, 1998, p. A37.
141. Jasper Becker, "Tightening the Noose on Parties," *South China Morning Post*, December 5, 1998, p. 2.
142. Gordon G. Chang, "Terrorists for Democracy, Part II," *China Brief* 3, no. 6 (March 25, 2003), www.russia.jamestown.org/pubs/

view/cwe_003_006_002.htm "CSN [China Support Network] Declares No Confidence in Peng Ming: One Stray Extremist [and] CFF: A Splinter Group," May 29, 2003, www.kusumi.com/china support.net/news64.htm.

143. Shulong Wang, "Du Shi Keyi Shi Ren Mingzhi—*Chaoji Daguo de Bengkui—Sulian Jieti Yuanyin Tansuo Tuijie*" [Studying History Can Make One Wise: Review of *The Collapse of a Super Power: Deep Analysis of the Reasons for the Soviet Disintegration*], *Dang Jian Daokan* [Constructing the Party Magazine] 1, no. 127 (Zhonggong Jiangsu Shengwei Zuzhi Bu, 2002), djdk.myetang.com/200201/20020131.htm.

144. Such an idea was apparently discussed by Chinese officials during a trip by members of the Carter Center to China in the late 1990s.

Chapter 8. Averting the Crisis

1. While U.S. NGOs, business, and universities have become active players in cooperative ventures to strengthen China's legal infrastructure, increase the capacity of China's environmental NGOs, and develop new market-based approaches to environmental protection, the U.S. government remains far behind the curve. Not for lack of trying but rather for lack of funding and opportunity, the central U.S. government agencies, including the Commerce Department, the State Department, the Environmental Protection Agency, and the Department of Agriculture, among others, remain hamstrung in their efforts to promote U.S. interests in China. The U.S. Department of Energy appears able to move ahead in areas such as clean coal technologies and research cooperation on nuclear energy, perhaps because these efforts directly benefit U.S. commercial interests.

2. The State Department's Democracy, Human Rights, and Rule of Law program has embraced the environment as one of its targets of assistance in China. And the U.S. embassy in Beijing has thrown all of its (limited) economic weight behind supporting environmental governance in China.

INDEX

Acheson, Dean, 49
acid rain, 18, 72, 190, 204, 289n74
Action for Green, 157–59
Agenda 21, 107, 135, 187–89, 193, 296–97n18
agriculture, 37, 61, 65, 71, 76–80, 119–20, 265. *See also* Cultural Revolution; Great Leap Forward
air pollution, 71–72, 235; Asia Pacific, 237–38; carbon dioxide, 183–84; health issues, 19, 72, 85, 89–90, 294n140; particulates, 72, 289n72, 294n140; sulfur dioxide, 72–73, 195–96, 223, 289n72, 294n140; transportation sector, 74–75. *See also* automobiles; coal
Air Pollution Prevention and Control Law, 104–5, 297n18, 298nn38–39
Alford, William, 101, 104, 195–96, 297n24
Angkor Wat, 180
anti-rightist campaign, 50, 138
Asian Development Bank (ADB), 62, 177, 189–90, 194–96, 198–99, 211
Asia Pacific, 11–12, 222–23, 236–39; environmental protection agencies, 239–40; foreign investment, 12–13, 262; forestry, 238–39; lessons for China, 247–49; nongovernmental organizations, 242–47; political changes, 251–52; politics of resource exploitation, 239–42; public health, 237–38
Asia-Pacific Economic Cooperation (APEC), 177, 203–4
Audley, John, 12–13
Austria, 231–32
automobiles, 104, 119, 203, 261, 265, 267, 290n86

backyard steel furnaces, 52–53
Baiyangding Lake, 68–69
Balayan, Zori, 250
banks, 116, 302n102
battery recycling, 164–65
Becker, Jasper, 52–53
Beijing, 106, 126, 161, 174; air quality, 9, 71–72, 75, 85, 192–93; migrant population, 81–82, 292n124
Belt, Tamara, 12
Bhagwati, Jagdish, 11
biodiversity, 65–66
Bi Yuan, 45
Bohemia, 223, 230

Bol, Peter, 34–35
Bo Xilai, 264
British Petroleum (BP), 194, 196–97,
    209, 314n38
Brown, Lester, 78
Buddhism, 30, 36, 55
building codes, 193–94
Bulgaria, 224, 225, 232, 234
bureaucracy, 100–101, 109. *See also*
    environmental protection bureaus
businesses, 241–42; environmental,
    160–61. *See also* multinationals

Cambodia, 180
campaign mentality, 17–18, 23, 57,
    121–27, 258
campaigns, 91–92, 268; afforestation,
    123; deforestation, 121–23; desertifi-
    cation control, 67, 123–24; environ-
    mental protection programs, 91–92;
    To Get Rich Is Glorious, 59; grain
    self-reliance, 77–78, 258; South-
    North Water Transfer Project,
    125–27; Three Rivers and Three
    Lakes, 124–25. *See also* Go West
    campaign; Great Leap Forward;
    Three Gorges Dam; West–East
    pipeline
Canadian International Development
    Agency (CIDA), 158, 161, 187
Cang Yuxiang, 4–5
capacity building, 193, 196–98,
    314n38
Carey, John, 13–14
Carson, Rachel, 142
censorship, 140–43
Center for Legal Assistance to Pollu-
    tion Victims, 115–16
Changzhou, 102
Charter 77, 230
Chen Hongmu, 44
Chernobyl nuclear disaster, 227–28
Chiang Kai-shek, 15
Chi Haotian, 214
China: cycles of social transformation,
    36–37, 56; exports and imports, 182,
    202–3; gross domestic product, 19,
    25, 61–62, 88, 107–8, 117–18, 259;
    income levels, 2, 118, 210; mortal-
    ity, 46, 53; political and economic

context, 14–17, 20–23; sovereignty,
    94, 98, 204; stability, 25–26, 83, 214,
    269; statistics, 24, 47, 121–23
China: Air, Land, and Water, 189
"China and the World in the
    Nineties," 186
China Central Television (CCTV),
    150–51, 162–64
China Council for International Coop-
    eration on Environment and Devel-
    opment (CCICED), 187, 189,
    194–95, 197–98, 312n17
China Daily, 166, 208
China Democratic Party (CDP),
    133–34, 255
China Development Union (CDU),
    254–55
China Environment Fund, 134–35,
    190
China Environment News, 123, 166,
    167–68, 212
China Forestry News, 167
China Green Students Forum, 166
China in the Year 2000, 141–42
China on the Edge (He), 141–42
China Petroleum and Chemical Cor-
    poration (SINOPEC), 194, 196–97
China Youth Daily, 8
Chinese Academy of Culture, 146–47
Chinese Academy of Sciences, 194
Chinese Academy of Social Sciences,
    5, 256
Chinese art, 28–29
Chinese Communist Party (CCP),
    25–26, 174–75, 222, 266; criticism
    of, 252–53, 255; economic reform
    and, 60–61; judiciary and, 112–13
Chinese People's Political Consulta-
    tive Conference, 102, 143, 146, 148
Chinese Preventive Medicine Journal,
    84
Chinese Society for Sustainable Devel-
    opment, 135
Chongqing, 106, 118, 193, 208
cities, 6–7, 18–19, 264, 299n60; air
    quality, 9, 71–72, 75, 192; Asia Pa-
    cific, 238; authority and, 117–21;
    cleanup efforts, 160–63; interna-
    tional aid to, 190–91; mayors, 109,
    117, 119, 260, 264–65; migrant pop-

ulation, 81–84; social unrest and, 61, 83–84; water supply, 68–70, 83, 88
Citigroup, 206–7
citizens, 56, 158; protests, 19, 45, 85–88, 113–15, 118, 227–28
Clean Development Mechanism (CDM), 185
cleaner production systems, 198–99, 314n43, 314n46
climate change, 65–66, 181–85, 261, 266–67, 296n16, 311n10, 312n12
Clinton, Bill, 133, 190, 192
coal, 19, 72–74, 104, 223, 289n74; heating units, 192–93; inspection teams, 109; market-based reforms, 100
coal mines, 67, 73, 289–90n76
Collapse of a Superpower, The, 256
"Communique on the State of China's Environment," 99
Communist Party, Russia, 232
Confucianism, 17, 27–33, 36, 55, 56
conservationists, 136–37, 145–49
constitution, Chinese, 95–96
Convention on International Trade in Endangered Species (CITES), 66, 153–54, 180
Copsa Mica (Romania), 233
cross-boundary pollution, 204–5, 230–32, 251
cultural artifacts, 208, 216
Cultural Revolution, 15, 53–55, 95, 138
Czechoslovakia, 223–25, 228, 230–34, 250

Dacheng Lane Neighborhood Committee, 161
Dachuan village, 86–87
Dadong Industries, 85–86
Dai Qing, 137, 142–47, 160, 162, 174
Dalian, 93, 106, 118–19, 164–65
Dalian Student Environmental Society, 167
dams, 2, 52, 204–5, 228, 231, 251. *See also* Three Gorges Dam
Danube River, 231–32, 251
Daughters of the Earth (Liao), 160
Daya Bay nuclear power plant, 170–71, 215–16

deforestation, 9–10, 18, 53, 64–67, 89, 116, 257, 286n16; campaigns, 44, 121–23; endangered species and, 149–56; flooding and, 9, 67, 121, 151; historic context, 40–45; illegal logging, 64, 122–23, 149–56, 158, 240–41, 243; imports, 122, 202–3, 238–39; logging bans, 121–22; western areas, 211–12
democracy, 129, 133–34, 137, 140–41, 169, 326n2; activists, 251–52, 254–55; Asia Pacific, 244–46; Eastern Europe, 250–51
Deng Nan, 188
Deng Xiaoping, 50, 61, 139; environmental protection efforts, 92, 95; legal system, 112, 127; reform process, 15–16, 59
Department of Energy (United States), 192
desertification, 9, 18, 40, 212, 257–58; campaigns, 67, 123–24; migration and, 66, 82
Desta, Asayehgn, 11, 13
developing versus developed countries, 93–94, 98–99, 183–84, 296n16
Dianchi Lake, 69, 124–25, 152
Dorje, Gisang Sonam, 153–54
Dorje, Zhawa, 153–54
droughts, 68–69, 78, 119
Duan Changchun, 158–59
Dubcek, Alexander, 250
Dunstan, Helen, 42, 43, 45–46
dynasties, 17, 29–32, 35–42, 45–47, 214, 279n13, 283n109

Earth Day, 157–58, 162, 173
Eastern Europe, 130–31, 141, 221–22, 223–24; enhancing government awareness, 224–26; environmental activism, 226–27, 230–31; future prospects, 233–36; political change, 249–51; social issues and, 229–33
ecological (re)construction, 118, 212, 213
economic development, 4–5, 25, 193–96, 228, 269; Asia Pacific, 236, 247–48; context, 14–17; debates, 10–14; demands on natural resources, 18, 62–63, 68, 72; Eastern

economic development (*continued*)
Europe, 234–35, 247–48; global implications, 25, 63, 73–75, 222; international cooperation and, 186–193;
population growth and, 75–77
Eichenberger, Joseph, 194–95
Elvin, Mark, 37, 41
emperors, 32, 283n109
endangered species, 66, 139, 145, 147,
149–56, 180–81; pandas, 64, 180–81;
Tibetan antelope, 149, 153–55; Yunnan snub-nosed monkey, 149–53, 155
energy sector, 183–84, 193–94
energy service companies (ESCOs),
199
Environment Agency (Japan), 240, 247
environmental activism, 22, 138–45;
Eastern Europe, 226–27, 230–31; future of, 173–75; media and, 163–66.
See also conservationists
Environmental Basic Law (Japan), 240
environmental crimes, 114–15
Environmental Defense (U.S.), 195
environmental degradation and pollution, 62–64, 215; cross-boundary,
204–5, 230–32, 251; economic costs
of, 8–9, 25, 60, 88–90, 184, 258; effect of philosophical views on,
36–41; legacy of, 17–23, 55–57; Mao
era, 47–55; national security and,
77–81; nineteenth century, 45–46;
positive outcomes, 19–20, 24;
trends, 23–25. *See also* air pollution;
water quality
environmental disasters, 9–10, 41, 93,
168, 227–28; Huai River, 1–9, 168.
See also droughts; flooding; population growth
environmental education, 146–48
environmental laws, 20, 97–98, 101–2;
drafting, 103–5; Eastern Europe,
224–25; enforcement and incentives,
199–201, 218; Legalist ideals, 30,
35–36, 47, 56–57, 92. *See also* Air
Pollution Prevention and Control
Law
environmental protection agencies,
Asia Pacific, 239–40
Environmental Protection and Natural

Resources Committee (EPNRC),
101, 104–5
environmental protection bureaus
(EPBs), 96–97, 105–12, 114–15,
147–48, 259–60, 302n106
Environmental Protection Law, 96–97
Environmental Protection Leading
Group, 95–96
environmental protection programs,
19–20, 63; fee system, 105, 109–12,
300n65; fines 114, 195, 299n59;
funding, 107–8; government role, 91,
97; historic context, 43–46,
281–82n70; institutions, 96–97; legal
system and, 97–98, 101–5; local officials and, 91–92, 95; market-based
approaches, 91–92, 100; personal initiatives, 91, 115, 117–21, 123–24,
140, 165; principles, 94, 98; public
participation in, 99; steps toward environmental governance, 93–98;
taxes 103; traditional approaches,
17, 23
Environmental Resources Management (ERM), 216
environmental responsibility system,
117–21
EP3 spy plane, 271, 273
ethnic minorities, 19, 76, 209, 214–15,
217, 267–68
European Union, 203, 234, 248
Export Import Bank (EX-IM Bank),
192, 207

factories, 1–6, 47, 51, 72–73, 120–21
Falun Dafa, 132, 134, 305n9
Falun Gong, 207, 253, 305n9
Fang Jue, 254
fertilizer pollution, 11, 71, 86, 125
Feshbach, Murray, 228
fisheries, 4, 88, 204
flooding, 9, 67, 121, 151, 255
Focus, 164
Foshan, 119–20
Four Modernizations, 15
Fourth Landmark, The (Peng), 254
Framework Convention on Climate
Change, 182
Friendly, Alfred, 228

Friends of Nature, 129, 136, 146–49, 151, 154, 157, 167, 307n47
Friends of the Earth, 170–72, 239, 309n73
Furtado, Jose, 12
future scenarios: environmental meltdown, 269–72; green, 264–67; inertia, 267–69
Fuyang, 6–7

Gabcikovo-Nagymaros Dam, 231
Gan Yuping, 208
General Motors (GM), 197, 199, 315–16n55
Geng Haiying, 137, 164–65
German Democratic Republic (GDR), 223, 225
global economy, 63–64, 73–75, 222
global environmental concerns, 25–26, 178–81; international cooperation, 185–93. *See also* climate change
Global Environmental Facility, 189, 191, 315n55
Global Greengrants Fund, 158
Global Village of Beijing, 137, 151, 160–63
Glueck, Jeffrey, 230, 250–51
Gorbachev, Mikhail, 250, 271
governance, 43; devolution of authority, 63, 117–21, 259–60, 268; Legalist period, 35–36; personalistic system, 17, 20–21; transformation of, 15–16. *See also* local officials
government-organized nongovernmental organizations (GONGOS), 99, 134–35, 146
Go West (Great Opening of the West) campaign, 81, 108, 148, 210–17; criticisms of, 211–13; future scenarios, 266–68, 270; national security and, 213–15
grain-growing campaigns, 19, 77–80, 258, 291n102
grassroots environmentalism, 156–60, 164–65
Great Leap Forward, 15, 51–53, 157
Green Camps, 139, 150–51, 158, 166
Green China, 146
green communities, 158, 161, 173

Green Culture Sub-Academy, 146–47
Green Earth Volunteers, 136, 163
Greenpeace, 146, 166, 168, 170–71, 239
Green Plateau Institute, 149, 152, 159, 166
Green Politics and Global Trade (Audley), 12–13
Green River Network, 136, 147, 154, 156
Green Times, 154, 166, 167
Green Volunteer League of Chongqing, 147, 157–58
Green Weekend, 167
Green World Tour, A (Tang), 139–41
groundwater, 7, 68–69
Guangdong Province, 73–74, 170–71, 291n104
Guangming Daily, 143
Guangzhou, 68, 118–20
Guangzhou Environmental Protection Office, 109–10
Guanting Reservoir, 93
Guan Zhong, 38, 42, 56
Guiqiu Convention, 38
Guizhou Province, 210
Guo Shuyan, 208

Hainan, 147–48
Hai River, 69
health issues, 19, 60, 72, 84–86, 88, 223, 257–58; air pollution, 72, 85, 89–90, 294n140; Asia Pacific, 237–38; birth defects and cancer, 2–4, 84, 86, 223; costs of pollution, 89–90
Hebei Province, 108–9, 195
He Bochuan, 137, 141–42
Hefei Iron and Steel Company, 198–99
Henan province, 43–44
He Qinglian, 255, 291n97
Hong Kong, 73–74, 170–72, 211, 237, 310n100
Hopewell Group, 171–72
hotlines, 118, 134, 165
household responsibility system, 61
Hou Yi, 29
Ho Wai-chi, 171
Huai River, 1–9, 69, 168

Huang, Ray, 34, 38
Hu Angang, 211
Huangpu River, 264
Hucker, Charles, 30, 33–34
Hu Jingcao, 138, 154–55, 164, 166, 168
Hu Jintao, 92, 127, 214, 274
Hu Kanping, 138, 154, 166, 167, 265
human rights, 245–46
Human Rights Watch, 207
Hungary, 223–24, 226, 231–32, 251
Hu Yaobang, 167

Indonesia, 239, 242–43
infrastructure projects, 17–18, 22, 52, 62, 194. *See also* Three Gorges Dam
Inglehart, Ronald, 11
inspection teams, 4, 106, 109–10
intellectuals, 22, 55, 138, 145, 179, 239, 250, 261
international community, 16, 25–26, 120, 161, 177–78, 260–62; climate change and, 181–85; enforcement and incentives, 199–201, 218; environmental aid, 21–22, 189–93; environmental cooperation, 179–80, 185–93; future scenarios, 266–67, 269; policy design and, 193–96; technology transfer, 198–99; trade regimes, 202–5; transformation of China and, 217–19
international governmental organizations (IGOs), 62, 100, 177, 194
international nongovernmental organizations (INGOs), 166, 168, 170, 177, 190, 194, 297n21
International Rivers Network, 166, 206–7, 232
International Standards Organization (ISO), 197, 241
Internet, 133, 136, 168, 255
iron and steel production, 52–53

Japan, 46, 120, 237–38, 240–42; acid rain, 72, 190, 204; environmental aid to China, 190–92; local initiatives, 241–42; nongovernmental organizations, 246–47
Jiande lawsuit, 113–14
Jiang Zemin, 10, 76, 78, 92, 126–27, 133, 156, 190
Jinchang Li, 38–40

judiciary, 92, 101, 112–17
Jun Jing, 86–88

Kekexili, 153
Kunming, 85–86, 124–25, 158
Kwong-to, Plato Yip, 171
Kyoto Protocol, 185, 261

lakes, 68–69, 71, 124–25, 152
Latvia, 229–30
lawsuits, 21, 102, 112–16, 301n82
lawyers, 113, 262
lead poisoning, 74, 85, 233
Lee, Yok-shiu, 12
Lees, Martin, 186
legal centers, nonprofit, 21
Legalism, 30, 35–36, 47, 56–57, 92
legal system, 24, 92, 163; environmental protection programs and, 97–98, 101–5; judiciary, 92, 101, 112–17; Mao's view, 47–48; post-Mao, 258–59. *See also* environmental laws
Liang Congjie, 99, 136, 146–49, 150, 164, 166, 173–74, 307n47; Tibetan antelope and, 154–55
Liang Qichao, 146
Liang Sicheng, 146
Liang Xi, 47
Liao River, 69
Liao Xiaoyi (Sheri Liao), 137, 152–53, 157, 160–63, 174, 213
Liebman, Benjamin, 104, 297n24
Li Li, 166, 265
Lin Feng, 143
Lin Kuang, 125
Li Peng, 4, 76–77, 187
Li Rui, 143–44
Li Si, 35
Liu Chuxin, 79
Liu Yan, 41
Li Xiaoping, 164
lobbying, 137, 170, 173
local officials, 4–5, 20–21, 23, 63, 102, 136, 260; environmental protection bureaus and, 108–11; environmental protection programs, 91–92, 95; environmental responsibility system, 117–21; judiciary and, 112–13; Malaysia, 240–41. *See also* governance

London Amendments, 181–82
London Convention against Ocean
  Dumping, 179–80
Lu Fuyuan, 191
Lu Zhi, 154
Lysenko, Trofim, 52

Ma Hong, 186
Ma Kai, 191
Malaysia, 122, 239–41, 242–43
Ma Xiaoying, 102, 118
management training, 196–97
Manning, Robert, 290n76
Mao era, 47–55, 77, 81; population
  growth, 49–51; propaganda, 52, 54
Mao Zedong, 2, 15, 75, 112, 125, 138,
  283n109; conception of nature,
  48–49; view of legal system, 47–48;
  view of population, 49–51
Marine Environmental Protection
  Law, 179
marine pollution, 179–80, 204
market-based reforms, 16, 59, 91–92,
  100, 119, 182, 194–96, 222
Ma Yinchu, 50
mayors, 109, 117, 119, 260, 264–65
Ma Zhong, 80
media, 130, 138, 143, 150–51, 166–68,
  247, 262; assistance to NGOs,
  154–55; environmental advocacy
  and, 163–66; environmentalist use
  of, 160–62; Mao era, 48–49, 52; television, 150–51, 160–64
Meili Snow Mountain, 149
Mekong Delta, 204–5
Mencius, 31–32, 38, 42
migration, 18–19, 37, 60–61, 81–84,
  270, 292n120; Beijing, 81–82,
  292n124; due to desertification, 66,
  82; harmonization and, 82–83; historic context, 39–40, 42; resettlement, 45, 52, 82, 126, 206–8,
  215–16, 244; social unrest and, 60,
  293n130; Three Gorges Dam project
  and, 126, 206–8
Minebea Group, 241–42
Ministry of Foreign Affairs, 183–84
Ministry of Foreign Trade and Economic Cooperation (MOFTEC), 191,
  202

Ministry of International Trade and In-
  dustry (Japan), 241–42
Ministry of Supervision, 106, 110
Ministry of Water Resources, 3, 68,
  105
Mogao Grottoes of Dunhuang, 180
Montreal Protocol, 182
Morgan Stanley, 206
multinationals, 12, 64, 197, 198–200,
  215–17, 262; international partner-
  ships, 196–97, 314n40
Murphey, Rhoads, 33, 48–49

Nagle, John, 101–2, 113
National Conference on Environmental Protection, 95, 132
National Environmental Protection
  Agency (NEPA), 51, 70, 100, 104,
  146,182, 297n25. *See also* State Environmental Protection Administration
National Environmental Protection
  Bureau, 96–97
nationalism, 251
Nationalists, 46
National People's Congress (NPC), 75,
  96, 100–101, 105, 206, 297n24
national security, 19, 23, 55, 77–81,
  213–15, 319n112
natural gas, 209–11
Natural Resources Defense Council,
  194
Nature, 139
nature reserves, 66, 140, 154, 165
Neo-Confucianism, 34–35
nongovernmental environmental advocates, 115–16
nongovernmental organizations
  (NGOs), 21–23, 99, 129–31, 209,
  252, 305n1; activist efforts, 136–38;
  Asia Pacific, 242–49; conservationists, 145–49; criticism of, 171–72;
  democratic, 133–34; Eastern Europe,
  229–32, 234–35; funding, 158, 163,
  247, 309n73; future prospects,
  262–63, 265, 268, 271–72; government and, 131–38; grassroots environmentalism, 156–60; Hong Kong,
  170–72; international, 166, 168, 170,
  177, 190, 194, 297n21; outside

nongovernmental organizations (*cont.*)
Bejing, 155–56; as political cover for
government, 173–74; registration,
133, 146–47, 152, 158–60; societal
needs and, 132–33; support for,
132–33. *See also* government-organized nongovernmental organizations
nuclear power, 170–71, 215–16,
227–28, 235

Oder River, 230–31
Olympics, 174, 192, 261, 267
Oriental Horizon, 154, 164
Ortolano, Leonard, 102, 118
Overseas Private Investment Corpora-
tion (OPIC), 192, 273
ozone depletion, 181–82, 185

Pak Mun hydropower project (Thailand), 244
pandas, 64, 180–81
Pearl River, 69, 73
Peng Ming, 254–55
People's Daily, 7–8, 78, 131
People's Liberation Army (PLA), 113,
157
personal initiatives, 17, 20–21, 56, 115,
140, 165, 262; desertification campaigns, 123–24; environmental responsibility system, 117–21
pesticide poisoning, 85
Peterson, D. J., 232
PetroChina, 209, 215–16, 253
Philippines, 236, 238–39, 244–45
philosophies, 30–36, 55–57; effect on
environmental degradation, 36–41.
See also traditional values
Poland, 223–25, 226–27, 228, 230–31,
234
policy design, 193–96
Polish Ecology Club, 226–27
political change: Asia Pacific, 251–52;
China, 252–56; Eastern Europe,
249–51
political participation, 99, 129–30,
222–23, 243–44
population growth, 10, 13–14, 19, 27,
60, 278n60; economic reforms and,

75–77; historic context, 39, 42–43;
Mao era, 49–51
Potemkin villages, 52
poverty, 77, 210, 213
Prison Memoirs and Other Writings
(Dai), 145
propaganda, 7–8, 52, 54, 93–94
public opinion, 104, 131, 163, 172, 225

Qu Geping, 38–40, 51, 96–97, 101,
105, 108, 191; international conference and, 186–87

rainfall, 68
*Relations of United States and China,
The* (Acheson), 49
religion, 132, 134, 252–53
"Rite of Zhou: Regional Officer," 29
river diversions, 67, 71, 88–89, 258.
See also South-North Water Trans-
fer Project
Romania, 224, 232–33
Rozelle, Scott, 118

Saich, Anthony, 134
sandstorms, 66, 180
sanitation, 71, 77, 83, 203, 236–37, 258
Schaller, George, 153
Schell, Orville, 153
scientific community, 182–84, 226
separatism, 19, 215
Seven-Li Trench, 6–7
severe acute respiratory syndrome
(SARS), 84, 255
Shandong Province, 1, 8, 110
Shanghai, 190–92, 258, 264, 292n124;
environmental protection efforts,
109, 111, 114, 117, 119–20; mayor,
109, 117, 119; water scarcity, 68, 83
Shanghai Automotive, 197, 199
Shang Yang, 41, 47–48
Shanxi Province, 109, 116
Shapiro, Judith, 48
Shcherbak, Yury, 250
Shell, 215–17, 314n38
Shi Lihong, 151–54, 166
Siberian River, 228
Sichuan Province, 10, 64–65, 67,
117–18, 121, 163

Silent Spring (Carson), 142

Smil, Vaclav, 11–12, 54–55

So, Alvin, 12

social organizations, 132–33

social unrest, 24–26, 207, 253, 270, 301n82; Eastern Europe, 229–33; environmental protests, 19, 85–88; migration and, 60, 293n130; political concerns, 132–34, 215, 223, 252–57; price hikes and, 119–20; United States and, 272–73; urban areas, 61, 83–84

social welfare, 16, 130

soil erosion, 43–44, 54–55, 212

Solidarity, 230

Songhua River, 69

Song Jian, 4, 132–33, 150, 186–88

Source of the Yangtze, The (Yang), 156

South Korea, 72–73, 204, 237, 238–39, 245

South-North Water Transfer Project, 89, 120, 125–27

Soviet Union, 221–24, 227–28, 256, 271. *See also* Eastern Europe

Sparks, Marcia, 139

Special Economic Zones, 62

State Council, 4, 95–96, 101, 105; Structural Reform Office, 107–8, 191

State Development and Planning Commission (SDPC), 107–8, 184, 188, 191, 299n50. *See also* State Planning Commission

State Development and Reform Commission (SDRC), 107–8, 110, 191, 194

State Economic and Trade Commission (SETC), 106, 108, 184, 191

State Environmental Protection Administration (SEPA), 7, 69, 72, 100, 148,189, 297n25; conservationists and, 147–48; environmental protection bureaus, 105–12; fee system, 110–12; inspection teams, 109–10; limitations, 20, 63, 213; media and, 167–68. *See also* National Environmental Protection Agency (NEPA)

State Environmental Protection Commission (SEPC), 3, 96, 101, 107, 189

State Forestry Administration, 105, 110, 122–23, 167

State Oceanic Administration, 74, 105, 180

state-owned enterprises (SOEs), 269–70; dismantling of, 16, 61, 83–84; pollution caused by, 63, 104, 110, 118, 222

State Planning Commission (SPC), 104, 107, 183, 188, 299n50. *See also* State Development and Planning Commission

State Science and Technology Commission (SSTC), 107, 188, 299n50

Su Kiasheng, 7–8, 21

Sumatra, 242–43

Sun Yat Sen, 205

Supreme People's Court, 100–101, 112

sustainable development, 99, 137, 188–89, 204, 296n17, 312n23

Taiwan, 15, 25, 73, 121, 179, 237, 239, 245–46

Taiyuan, 67–68

Tang Jiaxuan, 204

Tangshan, 85

Tang Xiyang, 137, 138–41, 145, 150, 159–60, 166, 173

Taoism, 30, 33–34, 55, 279–80n23

Tara River Dam, 228

technology transfer, 11–14, 182–84, 198–201, 218, 273, 315–16n55, 316n58

textiles, 203, 262

Thailand, 23, 238, 244, 247

Third National Conference on Environmental Protection, 97

Three Gorges Dam, 82, 118, 150, 205–9, 265; criticism of, 159, 162; Dai and, 143–44, 147, 162; international community and, 205–9; resettlement, 126, 206–8; water pollution, 208–9

Tiananmen Square, 140–42, 144, 167

Tian Dasheng, 147, 157–58

Tian Guirong, 165

Tibet, 81, 139, 149, 153, 209, 214–15, 217, 267

Tibetan antelope, 149, 153–55

Time for the Environment, 162
Tongda Rubber, 85
tourism, 87
township and village enterprises
    (TVEs), 16, 62–63, 104; pollution
    generated by, 70–71
toxic waste dumping, 74
Toyota Motor Corporation, 241
trade and foreign investment, 12–13,
    62, 262
trade regimes, 201, 202–5. *See also*
    Asia Pacific Economic Cooperation;
    World Trade Organization
traditional values, 17, 23, 27–30,
    55–56, 187. *See also* philosophies
Transcaucasus Main Railway, 232–33
transportation sector, 74–75
treaties and regimes, 178–85
Trillium Asset Management, 206–7

Ukraine, 227–28, 250
unemployment, 83, 84, 121–23
United Nations, 15
United Nations Conference on Envi-
    ronment and Development
    (UNCED), 98–100, 107, 162, 185–89,
    202, 261, 296n16, 296–97n18,
    297n21
United Nations Conference on the
    Human Environment (UNCHE), 20,
    93–95, 97–98, 178, 224, 261
United Nations Development Pro-
    gramme (UNDP), 107, 161, 191
United Nations Environment Pro-
    gramme, 97, 161, 180
United Nations Food and Agriculture
    Organization, 122–23
United Nations World Summit on
    Sustainable Development, 162,
    184–85, 261
United States, 28, 65, 66, 68, 74, 162,
    190, 206–7, 272–74, 326nn1–2
United States Agency for International
    Development (USAID), 192, 273

Vietnam, 204–5, 237

Wall, Derek, 28
Wang Canfa, 115–16

Wang Lixiong, 149
Wang Luqing, 167
Wang Shiduo, 42–43, 76
Wang Yongchen, 136, 163, 172
Warring States period, 31, 37–38
wastewater, 70, 84, 86, 114, 125,
    208–9, 288n58
Water Pollution Prevention and Con-
    trol Law, 101–2
water prices, 119–20
water quality, 18, 124–25, 203, 208–9,
    288n54, 288n58; Asia Pacific,
    236–37, 240; Daya Bay project,
    215–16; Eastern Europe, 223–24; en-
    vironmental laws, 101–2; fisheries,
    88, 204; groundwater, 7, 68–71; ma-
    rine pollution, 179–80, 204
water supply, 7–9, 67–71, 86, 257,
    287n35, 287n41; cities, 67–70, 83,
    88, 195; river diversion plans, 88–89,
    120, 125–27, 228, 258
Wei Haokai, 102–3
Weller, Robert, 34–35
Wen Bo, 138, 150–51, 165–67, 265
Wen Jiabao, 187, 264, 274
West–East pipeline, 194, 209–11,
    215–17, 265, 270
"Who Will Feed China?" (Brown), 78
Wild China, 136–37, 152
Wild Yak Brigade, 153–55
"Will Saving People Save Our Planet?"
    (Carey), 13
Worker's Daily, 7–8
World Bank, 317n80; environmental
    aid, 8, 62, 108, 162, 180, 189–91,
    315n54; environmental reports,
    67–68, 77, 85, 88, 110, 189, 198,
    200–201, 236–37; Go West campaign
    and, 211, 214–15
World Health Organization (WHO),
    72, 74, 85
World Trade Organization (WTO),
    75, 177, 202–3, 249, 261–62, 265,
    270
World Wildlife Fund (WWF), 151–52,
    154, 180–81
Wu Dengming, 157–58
Wu Fengshi, 135
Wuhan, 111

Xiao Zhouji, 212
Xie Zhenhua, 100, 106–9, 209, 212;
    economic development and, 63, 78;
    Huai River and, 4, 7; support for
    green communities, 161, 173
Xinjiang, 81, 214–15, 267
Xiuzhou, 6
Xi Zhinong, 136, 149–52, 154–55, 158
Xu Haiping, 125
Xu Kuangdi, 109, 119, 192

Yangtze River, 9, 67, 69, 121, 156, 208;
    flooding, 9, 151, 255
Yangtze! Yangtze! (Dai), 143–44, 307n41
Yang Xin, 136, 147, 154, 156–57, 172
Yan Jun, 150
Ye Jianying, 144
Yellow River, 52, 69
Ye Shi, 41
Ying-sheung, Gordon Wu, 171
Yu Gong, 49

Yugoslavia, 224, 228
Yungang Grottoes, 180
Yunnan Environmental Science Soci-
    ety, 159
Yunnan Province, 44–45, 213
Yunnan snub-nosed monkey, 149–53,
    155
Yu Sen, 43–44, 56

Zeng Peiyan, 212
Zero Hour Operation, 5–6
Zhang Guangduo, 208
Zhangjiakou, 195
Zhejiang Province, 52, 113–14, 118
Zhongnanhai, 253
Zhongshan, 119, 120–21
Zhou Enlai, 15, 75, 93–95, 97
Zhu Rongji, 61, 66–67, 71, 106, 121,
    151, 164, 184, 207
zinc mines, 84
Zvosec, Christine, 228